EVOLUTIONARY STRATEGIES OF PARASITIC INSECTS AND MITES

EVOLUTIONARY STRATEGIES OF PARASITIC INSECTS AND MITES

Edited by
Peter W. Price
Department of Entomology
University of Illinois
Urbana, Illinois

PLENUM PRESS · NEW YORK AND LONDON

Library of Congress Cataloging in Publication Data

Symposium on Evolutionary Strategies of Parasitic Insects and Mites, Minneapolis,
   1974.
   Evolutionary strategies of parasitic insects and mites.

   Held at the national meeting of the Entomological Society of America.
   Includes bibliographies and indexes.
   1. Insects—Evolution—Congresses. 2. Mites—Congresses. 3. Arachnida—Evolu-
tion—Congresses. 4. Parasites—Congresses. I. Price, Peter W. II. Entomological
Society of America. III. Title.
QL468.7.S92   1974   595.7′05′24   75-11524
ISBN 0-306-30851-7

Proceedings of a symposium on Evolutionary Strategies of Parasitic Insects
and Mites held at the national meeting of the Entomological Society of America
in Minneapolis, Minnesota, December 2-5, 1974

© 1975 Plenum Press, New York
A Division of Plenum Publishing Corporation
227 West 17th Street, New York, N.Y. 10011

United Kingdom edition published by Plenum Press, London
A Division of Plenum Publishing Company, Ltd.
Davis House (4th Floor), 8 Scrubs Lane, Harlesden, London, NW10 6SE, England

Printed in the United States of America

PREFACE

This volume contains the invited lectures presented in a symposium entitled "Evolutionary strategies of parasitic insects and mites" at the national meeting of the Entomological Society of America in Minneapolis, Minnesota, 2-5 December, 1974. The intent was to bring together biologists who have worked on arthropods that are either plant or animal parasites in order to foster consideration of general aspects of the parasitic way of life. There seems to be a deficiency of ecological and evolutionary concepts relating to parasitism, in contrast to the burgeoning literature on predation, and it appeared that an amalgamation of studies on plant and animal parasites might help development of some generalities. Since parasities are far more numerous than predators in the world fauna, or in any particular community, emphasis on their study is justified. I freely admit that parasitoids have been usefully regarded as predators by ecologists, and many concepts on predation have been derived from their study. Also, in whichever category one places the parasitoids, that is the one which contains the most species. However, from an evolutionary point of view they show many characteristics that must be regarded as those of a parasite. Notably, they are small, highly specific to their host, highly coevolved with it, as a result many species can coexist, and their adaptive radiation has produced the majority of the species diversity seen on Earth today.

For such rapidly evolving organisms as parasites which must certainly adapt to the changing nature and resistance of the host, an evolutionary approach to their study seems particularly relevant. Therefore this symposium has brought together evolutionary biologists that have in common an interest in parasitic arthropods. We try to observe the perpetual motion of the evolutionary process by studying a mere instant in evolutionary time. With this grave disadvantage we can gain only a crude picture of the molding influences on the evolution of parasites, but one that is nevertheless fascinating and, I hope, instructive.

This symposium would not have been organized without the publication of Dr. Askew's book entitled "Parasitic Insects" which has provided tremendous impetus to the field by integrating for the first time information on parasites of interest to the large disciplines of medical entomology and biological control. While reviewing the book it struck me that further integration might be beneficial. For example only some of the superfamily Cynipoidea were considered in the parasitic Hymenoptera. These were the parasites on other insects whereas the remainder of the members are nonetheless parasitic, but on plants rather than animals. It was possible that some evolutionary insight might be gained by using a comparative approach to both types of parasitism. Then when Dr. Bush visited the University of Illinois it became clear to us that there was much to share between workers on plant and animal parasites. We decided that a symposium would be valuable, and I was most fortunate that the original plan for it came to fruition since all the invited speakers agreed to participate. I am very grateful to them all. I am also grateful to officers of the Entomological Society of America, Drs. W. P. Murdoch, H. C. Chiang, A. C. Hodson, and F. W. Stehr for support of many sorts in the organization of the symposium. I must also thank Judy Michael for the precision with which she typed the final draft of the symposium, and Mr. Roy Baker, Managing Editor of Plenum Publishing Company, for making publication possible.

Urbana, Illinois   P.W.P.
January 1975

CONTRIBUTORS

R. R. Askew
    Department of Zoology, University of Manchester, England

G. L. Bush
    Department of Zoology, University of Texas, Austin, Texas
    78712, U.S.A.

D. C. Force
    Department of Biological Science, California State Polytechnic
    University, Pomona, California 91768, U.S.A.

D. H. Janzen
    Department of Biology, University of Michigan, Ann Arbor,
    Michigan 48104, U.S.A.

R. W. Matthews
    Department of Entomology, University of Georgia, Athens,
    Georgia 30602, U.S.A.

R. Mitchell
    Zoology Department, Ohio State University, Columbus, Ohio
    43210, U.S.A.

P. W. Price
    Department of Entomology, University of Illinois, Urbana,
    Illinois 61801, U.S.A.

S. B. Vinson
    Department of Entomology, Texas A&M University, College
    Station, Texas 77840, U.S.A.

CONTENTS

INTRODUCTION: THE PARASITIC WAY OF LIFE AND ITS CONSEQUENCES

Peter W. Price

Department of Entomology, University of Illinois

Urbana, Illinois 61801, U.S.A.

The parasitic mode of obtaining food is a major life strategy
among arthropods. Insects which are parasitic on animals constitute
about 10 percent of named species in the Animal Kingdom (Askew 1971).
For many large groups perhaps only 25% of the world fauna has been
described (e.g., Townes 1969, on Ichneumonidae, Matthews 1974, on
Braconidae), so percentage representation in the Kingdom should be
more than doubled. Insects and mites which are parasitic on plants
are even more numerous. Therefore probably at least half of the
animals on Earth are parasitic arthropods.

Parasitic arthropods are of interest to the majority of ento-
mologists. Those arthropods that are parasitic on plants are often
economic pests, although some are used in the biological control of
weeds. Those which parasitize higher animals include insects of
interest to the medical entomologist, and others which attack insects are important in biological control. These disciplines of
Agricultural and Forest Entomology, Medical Entomology, and Biological Control have tended to treat the process of parasitism in
different ways, with little flow of information or general concepts
among them. And yet it is clear that a leafhopper (Cicadellidae)
or a froghopper (Cercopidae) is just as much a parasite as a body
louse (Pediculidae) or a louse fly (Hippoboscidae). Also seed
beetles (Bruchidae) and seed chalcids (e.g., Megastigminae) are
parasitoids just as are tachinid flies (Tachinidae) and fairyflies
(Mymaridae) (see also Janzen, this volume). Parasites on plants
and animals share many problems in their life style, and many adaptive answers to these problems show convergent evolution in widely
divergent taxonomic groups. The common problems in the parasitic
way of life have evolutionary consequences both on the parasite and
on its host. These will be examined repeatedly in this book.

1

APPROACHES TO THE STUDY OF PARASITIC ARTHROPODS

A parasite can be defined broadly as an organism which lives
in or on another living organism, obtaining from it part or all of
its organic nutriment, usually having a negative influence on the
fitness of its host, and commonly exhibiting some degree of adaptive
modification. It is not always useful to think in such broad terms,
but frequently it is very instructive. However, depending upon the
interests and biases of the observer this large and diverse group
of organisms may be categorized in several ways: 1. An ecologist
might be most interested in the impact of the parasite on the host
individual and the host population; 2. An evolutionary biologist
would also be interested in their innate characteristics, the
adaptive nature of these and the patterns within and between taxonomic groups and ecological guilds; 3. A physiologist may be most
interested in the biochemical details of the host-parasite relationship, defensive mechanisms of hosts, and parasite and host development.

From these various vantage points has grown a jargon that is
much debated, particularly in relation to parasitic insects that
have a free-living adult stage. Ecologists have recognized that
the free living adults have an impact on the host population which
is identical to a *predator* (see also Janzen, this volume). Each
host on which an egg is deposited is killed by the parasitic larva,
thus the adult is responsible for killing many hosts just as is a
predator. Reuter (1913) recognized these facts and called these
"parasite-like predators" the Parasitoidea, from which the word
parasitoid has been derived. Since the basis for the word is equivalent to a superfamily name we should not try, nor expect, to form
a verb from it, as some would have us do. The term has gained wide
usage and will no doubt outlive the newer terms of predatoid and
carnivoroid (cf. Flanders 1973) that have been suggested. However
the parasitoid also meets the general definition of a parasite
given above and those that have concentrated on the details of the
host-parasite relationship may find more merit in the term *parasite*
than in others.

From an ecological point of view the only fundamental differences that are necessarily recognized are whether the organism *saps*
energy from its host or whether it *slays* its host or prey. These
two basic life strategies are treated by Mitchell (this volume) as
m and *l* strategies respectively. An *m* strategist influences the
number of progeny produced by its host, whereas an *l* strategist
affects the age-specific survival of its host. But even this di-
chotomy is not absolute. As Vinson (this volume) emphasizes, the
parasitoid egg and larva may derive nutriment and live with the living host for a considerable time, so many adaptations typical of a
sapper are evident, yet ultimately the parasitoid becomes a *slayer*.
Also parasitic castration is a common phenomenon (e.g., Wheeler

1910, Dogiel 1964). When complete the host has zero fitness, and
apart from acting as a competitor with conspecifics for food or as
food for predators, it can be regarded as non-operational or dead.
Thus sappers may have an effect on the population dynamics of the
host similar to that of predators on prey, either through parasitic
castration, reduced production of gametes, inferior competitive
ability during sexual selection (cf. Wheeler 1910), or food acqui-
sition.

The evolutionary biologist who is interested in innate characteristics of organisms and their adaptive qualities may take a
comparative approach in order to understand the strategies resulting
from the evolutionary process for overcoming a common problem. One
such problem is whether a female parasite should invest much energy
per individual and produce few progeny, or whether many progeny with
little energy per individual is a better strategem. These contrasting K and r strategies are discussed in detail in this volume (Force,
Askew) and patterns may be seen in these strategies among parasite
species (Price), and in the community organization of parasite-host
communities (Force, Askew). In turn the host evolves strategies
in response to the heavy selective pressure from parasites. These
coevolutionary interactions are examined by Vinson and Janzen in
this volume. The comparative approach may also focus on behavioral
attributes of parasites considered in this volume which are important as isolating mechanisms (Bush, Matthews), and which therefore
promote speciation (Bush).

Whatever the bias of the observer, and whatever name one wishes
to use for organisms on the parasite to predator continuum, all
approaches must be employed to gain a full understanding of any one
organism. In this volume several approaches are examined, but the
need for integration of these approaches can be appreciated when the
process of coevolution (Vinson, Janzen) and speciation (Bush) are
considered.

EVOLUTIONARY CONSEQUENCES OF THE PARASITIC HABIT

Consequences for the Parasite

There are many factors involved in the evolution of the im-
pressive diversity of parasitic arthropods. Parasites are small
and of necessity specialized for living in or on the host. Parasite
niches are narrow. Since each host changes through its life cycle,
and even at any one stage a diverse array of niches may be available,
several to many parasite species can exploit a single host species.
Thus during the adaptive radiation of hosts large numbers of niches
rapidly become available for parasitic organisms, and these in turn
may be parasitized. Thus the evolution of one plant species may

open niches for ten to even 1000 parasitic organisms. The oaks
(*Quercus robur* and *Q. petraea*) in Britain sustain 284 species of
parasites (herbivores) (Southwood 1961) and some of the many niches
available for colonization are discussed by Morris (1974). These
herbivores must surely sustain a much larger number of their own
parasites than the trees, and these in turn support hyperparasites.
The complexity of one small part of this system is described by
Askew (this volume). Thus evolution of the two oaks, the long time
available for colonization by parasites, and oak abundance which
increases the probability of colonization, has resulted in an
increase in the number of species by more than two orders of magnitude. The adaptive radiation of the Leguminosae led to a correlated radiation of bruchid beetles and their parasites (see Janzen,
this volume). The adaptive radiation of pines in North America
was paralleled by the radiation of bark beetles, for example of the
genus *Ips*, which has in turn permitted the radiation of mites in
the genus *Iponemus* which are phoretic on adult *Ips* and parasitic
on their eggs (Lindquist 1969). Ehrlich and Raven (1964) suggested
that the adaptive radiation of butterflies (Papilionoidea) occurred
as new groups of plants evolved, and "reciprocal selective responses
between ecologically closely linked organisms" may have resulted in
much of the species diversity in terrestrial ecosystems. As might
be expected, the larger the family of hosts, the more species of
any one parasite taxon may be found attacking members of that family.
For example, when the numbers of agromyzid flies which attack a
certain family of plants and the size of the plant family are compared, the linear regression accounts for 61% of the variation in
numbers of agromyzid species per plant family (data on Agromyzidae
in Spencer 1969, on plant families in Fernald 1950).

     The coevolutionary interactions between parasite and host are
vital to the maintenance of both. The host must have certain defenses against the parasite which prevent uninhibited exploitation,
and the parasite must evolve to cope with, or to subdue these defenses. An equilibrium is necessarily developed between host and
parasite individuals, and as a result between their populations.
Disequilibrium may result where this coevolution has not developed
and either one or both antagonists may go extinct. Such a situation
may have been glimpsed by Zimmerman (1958, 1960) in Hawaii where
five species of pyraustid moths, *Hedylepta*, evolved and were abundant
in association with three endemic parasite species. However with
the introduction of 11 alien parasite species to control agricultural pests the species of *Hedylepta* became rare or extinct as a
result of extremely heavy parasite pressure. No doubt many of the
successes and failures in biological control have been due to the
degree of coevolution that had been possible before introduction
of the control agent, or the degree to which the coevolved systems
had degenerated during allopatry.

     Coevolution results in increased specialization of the parasite, and the more highly adapted individuals utilizing a certain strategy must be, the narrower will be their niche exploitation, with the potential result that more species can evolve and coexist. For example, in the British fauna there are increasing numbers of species per family as the parasitic habit demands more specialization. Considering only the largest families, sizes range from the Miridae (187 species), Cicadellidae (265 sp) and Aphididae (366 sp) which feed externally on plants, to the Curculionidae (524 sp) and Cecicomyiidae (629 sp), where the larvae feed internally on plant material, and finally to the Braconidae (811 sp) and Ichneumonidae (1988 sp) the species of which live in or on an animal host which can be more active than a plant in its own defense. Many families with species that require similar degrees of specialization are much smaller but it appears that the upper limit to the number of species is closely related to the need for coevolutionary adaptations (see Table 1 for details). Indeed the comparatively small numbers of species per family (maximum = 46) in which most members are predators (see Table 1) suggests that the reciprocal selective pressures between parasitic herbivores of any type and their hosts have produced an impressive amount of diversity. Conversely the generalist tendencies of predators must suppress adaptive radiation relatively early since niches are necessarily broad and communities become rapidly saturated.

The very large size of many families of parasites suggests that speciation must be relatively rapid. The large number of families of parasitic insects also shows that adaptive radiation has taken many different courses compared to the rather limited scope of radiation seen among predators. Zimmerman (1958, 1960) presents evidence that five or more species of *Hedylepta* in Hawaii have evolved rapidly, since the banana was introduced 1000 years ago by man. Bush (this volume) presents evidence that speciation can be extremely rapid among phytophagous parasites. Askew (1968) has made a strong case for rapid speciation in the Chalcidoidea. He points out that all species are small and many develop gregariously. Sib mating is common. Even where parasites develop singly sibs are likely to mate since a female may deposit several eggs in close proximity. Thus gene flow in a population is restricted and segments may become isolated sufficiently for speciation to occur. This is particularly likely where an isolated pocket of hosts is colonized by a single female parasite. The parasite is already inseminated and her progeny are ensured of mates. Askew describes several other factors which may be applicable to parasitic arthropods in general. He further suggests that in the parasitic Hymenoptera, where the haplo-diploid form of sex determination results in haploid males, in each generation all the results of gene action are exposed, "unfavorable recessives can be eliminated and favorable gene combinations more rapidly selected." Thus a rapid evolutionary rate is probable.

Table 1. The number of species per family of insects in the
British fauna given in the check list by Kloet and Hincks (1945).
The families are categorized according to the ecological role of
members as predators, primary parasites (herbivores), and secondary
parasites (carnivores) (see Janzen, this volume, for development
of the concept that some insect herbivores may be regarded as
parasites, and therefore their parasites become hyperparasites, or
secondary parasites). Only some of the larger families are listed,
in which most of the species can be categorized with some certainty.

| Predators* | | Primary Parasites | | Secondary Parasites | |
Family	Number of species	Family	Number of species	Family	Number of species
Aradidae	4	Coreidae	25	Braconidae	811
Reduviidae	6	Tingidae	24	Aphidiidae	84
Nabidae	12	Miridae	187	Ichneumonidae	1988
Saldidae	20	Cicadellidae	265	Eurytomidae	77[†]
Anthocoridae	25	Delphacidae	68	Torymidae	89[†]
Hemerobiidae	29	Psyllidae	82	Encyrtidae	145
Chrysopidae	15	Aphididae	366	Pteromalidae	495
Coccinellidae	46	Coccidae	165	Eulophidae	486
Cucujidae	32	Cerambycidae	68	Tachinidae	234
Vespidae	27	Chrysomelidae	257		
Rhagionidae	20	Curculionidae	524		
Asilidae	26	Tenthredinidae	361		
		Cynipidae	263[††]		
		Cecidomyiidae	629		
		Tephritidae	75		
		Agromyzidae	95		
		Chloropidae	87		

*The families Empididae (327 sp) and Dolichopodidae (260 sp)
contain members that are predaceous as adults and some larvae
are also known to be predaceous. However not enough is known of
larval habits to place the families in this group. Indeed, the
large sizes of the families could suggest a scavenging habit of
many species also seen in other large families such as the
Carabidae (287 sp) and Staphylinidae (968 sp).

[†]Some species are primary parasites.

[††]Some species are secondary parasites.

The common problems that parasites are exposed to have been solved in a similar way repeatedly. These convergent strategies involve host searching, host selection, mating systems, protection of progeny and many others. For example, small organisms searching in a complex environment have evolved to respond to simple cues (for details see Vinson, this volume). Both primary and secondary parasites may utilize similar cues for host finding. The cabbage aphid, *Brevicoryne brassicae*, uses a component of mustard oil as a chemical cue for host discovery, and its parasite, *Diaeretiella rapae*, is also attracted to the aphid host by mustard oil (Read et al. 1970). Regarding mating systems, it is common for parasitic species to mate before dispersal. Again, mate discovery in a complex environment where species are small may be difficult, but the chance of founding new "colonies" is greatly increased by dispersal of inseminated females. Askew (1968) stresses the evolutionary importance of mating shortly after emergence. Matthews (this volume) describes the details of behavior in *Melittobia* species. Mitchell (1970) has shown in tetranychid mites and others how the probability of establishment at new sites is greatly increased by dispersal of inseminated females. In contrast, Bush (this volume) explains the important consequences of mating at the oviposition site for other parasitic species.

Another common problem is the defense of vulnerable eggs or larvae that are placed in or on a food source. This food might be utilized by members of the same or other species and the parasitic larva might be robbed of its nutriment and itself consumed. However, we see a convergent evolutionary strategy for female parasites to protect their progeny by the chemical defense of a territory. Females thus defend sufficient food on which their progeny can survive. Members of the large orders Coleoptera, Diptera and Hymenoptera have adopted the same strategy, and parasites of both plants and animals are involved. For example females of the apple maggot, *Rhagoletis pomonella* (Diptera: Tephritidae), chemically mark a territory which is probably large enough for a larva to mature within the fruit (Prokopy 1972). Females of the Azuki bean weevil, *Callosobruchus chinensis* (Coleoptera: Bruchidae) mark beans in which they oviposit (Oshima et al. 1973). Females of parasitic Hymenoptera show the same type of larval defense by marking hosts. Examples can be found in the Braconidae (e.g., Vinson and Guillot 1972), the Ichneumonidae (e.g., Price 1970, 1972) and the Chalcidoidea (e.g., Salt 1937, Askew 1968). We can infer from the few species that have been studied and the many taxa that are represented that this strategy has evolved in thousands of species.

Consequences for the Host

The effect of parasites on their host populations is almost certainly non-random in every case. Thus parasites act as agents

in natural selection causing evolutionary change in the host species.
This important influence of parasites has been almost ignored in much
of the scientific literature on parasites. We know very little about
qualitative differences among potential host individuals of the same
species that make them susceptible to, relatively protected, or immune
from parasitism. Therefore it is not possible to say in which evolutionary direction the parasite influence is moving the host population. Some examples will show how the parasitic way of life may
result in change of host distribution, behavior, breeding site construction, sociality, morphology and emergence time. These will be
treated in order to emphasize the importance of parasites as selective agents in the evolution of their hosts.

Distribution of the host may be influenced in several ways by
parasites. Parasites may be much more influential in one situation
than another so the plant host occupies a refuge from parasites.
Such a situation is seen in the klamath weed, *Hypericum perforatum*,
which now grows mostly in shaded sites where the chrysomelid beetle,
Chrysolina quadrigemina, lays few eggs (Huffaker and Kennett 1959).
In open sites the parasite is so successful in finding and eating
its host that both are rare. As Huffaker (1964) and Harper (1969)
point out, ecologists without a knowledge of the historical details
would probably describe the weed as a shade dwelling species and
the parasite as a rare and unimportant herbivore. We can only wonder
how many other distributions are so profoundly influenced by parasite
pressure. Cornell (1974) has suggested that a gap between geographic
distributions of allopatric species may be maintained by parasite
pressure provided that the species are closely related and the hosts
are more susceptible to the other species' parasite. Each host and
its habitual parasite may be coevolved for mutual tolerance but when
infection by the relative's parasite occurs the attack may be serious
or lethal so that an empty zone appears between the two species.
Barbehenn (1969) has also suggested that coexistence of two species
may be permitted if the weaker competitor carries parasites that
are more deleterious to its competitor than to itself. Food for
seed parasites is aggregated, and parasites cluster to the parent
plants. Thus dispersal of a seed to the perimeter of the seed shadow or beyond increases the chance of survival while also increasing
the distance between conspecifics. Janzen (1970, see also this
volume) suggested that this extensive spacing due to parasite (seed
predator) pressure maintains space for other species with similar
strictures on spacing, and a high species diversity is the result.
Thus parasites may have considerable impact on the ecology of individual species, community organization, and biogeography.

Parasites may also infest individuals with certain behaviors
more readily than others. Such a case has been witnessed among
rabbits in Britain after the introduction of a virulent myxoma virus
carried by the rabbit flea, *Spilopsyllus cuniculi*. The flea oviposits in nest material and progeny are ensured of finding hosts

since both are confined by the subterranean tunnels of the breeding site. Therefore the fleas rapidly transmitted the virus from one rabbit to another and since the virus was usually lethal there was, among other things, strong selection for reduced efficacy of the vector (cf. Rothschild 1961). In fact once the epidemic was over it was found that survivors were breeding above ground where the chance of contracting an infected flea was greatly reduced. Bark beetles (Scolytidae) also breed in tunnels and in this case, when under parasite pressure, there is little alternative but to continue doing so. However, Lindquist (1969) has suggested that those species of *Ips* which have *Iponemus* mite egg parasites reduce the chance of secondary infections of mites by not constructing ventilation holes along the brood gallery. These are a common feature in galleries of other species. Thus the only possible entrance for a mite is the gallery entrance which is frequently guarded by a male beetle. Indeed, it seems eminently reasonable that parasite pressure has selected for sociality in some insects (Michener 1958, Lin and Michener 1972). Where more than one bee occupies a nest one can act as a guard to the nest opening and block entry by parasites such as mutillids. Lin (1964) described the vigorous battles between guard bees and mutillids and how, when one guard failed, there would frequently be a second to plug the entrance. Under the influence of such aggressive parasites sociality must be strongly reinforced.

Where parasites search visually they may exert selection pressure for divergence of form among species of sympatric hosts so cross-recognition is reduced. Divergence of chemicals may also occur where these are used as cues by searching parasites. There is no reason to doubt the possibility that polymorphism of various sorts may also be adaptive against parasites, although I am not aware of any examples that result directly from parasitic pressure. However, Gilbert (1975) has suggested that the diversity of leaf shapes among sympatric *Passiflora* species has evolved in response to heavy herbivore pressure from *Heliconius* larvae. The visually searching and intelligent adult which learns to search for a particular leaf shape is thus unlikely to oviposit on congeneric species. Also the number of *Passiflora* species that are able to coexist may well be determined by the number of distinct leaf images that will not be confused by the *Heliconius* butterfly. Another remarkable feature of *Passiflora* morphology is the existence of extrafloral nectaries that attract secondary parasites (Trichogrammatidae) which oviposit in the eggs of the primary parasite, *Heliconius* (Gilbert 1975).

Finally many species of parasite probably affect the phenology of their host by attacking early or late individuals more severely. One example concerns the parasites of cocooned sawflies that overwinter in the leaf litter (Price and Tripp 1972). Those cocoons that lay in the upper litter were heavily attacked but those that were deeply buried were immune from parasites. Since the upper layers of the litter warm up first in the spring the sawflies at

these levels emerge first. Thus in effect the parasites killed
early-emergers, and the phenology of the host was altered to a later
mean emergence date. This change no doubt had further ramifications
throughout the life cycle of the sawfly.

CONCLUSIONS

The very nature of the parasitic habit sets certain restraints
on parasite evolution (see also Mitchell, this volume) so that many
species evolve the same solutions to common problems (see also
Vinson, this volume). Because of these restraints, patterns can be
observed in the relative abundance of parasitic species when compared to predators or in the degree of specialization required in
the parasite-host relationship. This also results in patterns seen
in the number of parasites per host, the number of parasites on each
stage of the host, and the community organization of parasites (see
Force, Askew and Price, this volume). In turn, the distribution and
abundance of hosts sets other limits on the parasites, and patterns
may be seen again (Janzen, this volume). Finally, the interaction
of host and parasite characteristics influence the way in which
species evolve (see Bush, this volume) and how reproductive isolation
is maintained between species (see Matthews, this volume). Parasites
also influence the host in many ways, but their full selective impact has yet to be appreciated.

The practical application of knowledge on the patterns seen in
the parasitic way of life should be readily evident. Agricultural-
ists must be aware of the potential for rapid shifts in host specificity of crop parasites and the strictures that they can impose on
this shift. They must also be able to manipulate beneficial parasites by means of pheromones and kairomones to improve impact on
pest species. For biological control purposes parasites can be
selected according to the stage in succession, the stage of parasite community development, or the degree to which the community
in which they must function is saturated. They may be chosen for
high colonizing ability or high competitive prowess with the knowledge that for each strategy there are many correlated attributes
which should be considered. In every case parasites act in a non-
random way and through the selection process coevolution results
in a more stabilized system. Efforts might be made to introduce
complementary pairs of parasites that counteract the directional
nature of each other's selective impact. However, much more study
on natural selection by parasites on their hosts is probably required
before practical use can be made of this aspect of the parasite-host
relationship.

ACKNOWLEDGEMENTS

I am grateful to all participants in this symposium for helping to bring into focus many attributes of parasites. I thank M. Mayse and J. Thompson for reviewing the manuscript and I appreciate financial support from U. S. Public Health Training Grant PH GM 1076.

LITERATURE CITED

Askew, R. R. 1968. Considerations on speciation in Chalcidoidea (Hymenoptera). Evolution 22:642-645.

Askew, R. R. 1971. Parasitic insects. American Elsevier, New York. 316 p.

Barbehenn, K. R. 1969. Host-parasite relationships and species diversity in mammals: an hypothesis. Biotropica 1:29-35.

Cornell, H. 1974. Parasitism and distributional gaps between allopatric species. Amer. Natur. 108:880-883.

Dogiel, V. A. 1964. General parasitology. Oliver and Boyd, London. 516 p.

Ehrlich, P. R., and P. H. Raven. 1964. Butterflies and plants: a study in coevolution. Evolution 18:586-608.

Fernald, M. L. 1950. Gray's manual of botany. 8th ed. American Book Co., New York. 1632 p.

Flanders, S. E. 1973. Particularities of diverse egg deposition phenomena characterizing carnivoroid Hymenoptera (with morphological and physiological correlations). Can. Entomol. 105:1175-1187.

Gilbert, L. E. 1975. Ecological consequences of a coevolved mutualism between butterflies and plants. p. 212-240. In L. E. Gilbert and P. H. Raven, eds. Animal plant coevolution. Univ. Texas Press, Austin.

Harper, J. L. 1969. The role of predation in vegetational diversity. Brookhaven Symp. Biol. 22:48-61.

Huffaker, C. B. 1964. Fundamentals of biological weed control. p. 631-649. In P. DeBach, ed. Biological control of insect pests and weeds. Reinhold, New York.

Huffaker, C. B., and C. E. Kennett. 1959. A ten-year study of vegetational changes associated with biological control of Klamath weed. J. Range Manage. 12:69-82.

Janzen, D. H. 1970. Herbivores and the number of tree species in tropical forests. Amer. Natur. 104:501-528.

Kloet, G. S., and W. D. Hincks. 1945. A check list of British insects. Kloet and Hincks, Stockport. 483 p.

Lin, N. 1964. Increased parasite pressure as a major factor in the evolution of social behavior in halictine bees. Insectes Sociaux 11:187-192.

Lin, N., and C. D. Michener. 1972. Evolution of sociality in insects. Quart. Rev. Biol. 47:131-159.

Lindquist, E. E. 1969. Review of holarctic tarsonemid mites

(Acarina: Prostigmata) parasitizing eggs of ipine bark beetles.
Mem. Entomol. Soc. Can. 60:1-111.

Matthews, R. W. 1974. Biology of Braconidae. Annu. Rev. Entomol.
19:15-32.

Michener, C. D. 1958. The evolution of social behavior in bees.
Proc. 10th Int. Congr. Entomol., Montreal 2:441-447.

Mitchell, R. 1970. An analysis of dispersal in mites. Amer.
Natur. 104:425-431.

Morris, M. G. 1974. Oak as a habitat for insect life. p. 274-297.
In M. G. Morris and F. H. Perring, eds. The British oak.
Classey, Faringdon, Berks.

Oshima, K., H. Honda, and I. Yamamoto. 1973. Isolation of an
oviposition marker from Azuki bean weevil, *Callosobruchus
chinensis*. Agr. Biol. Chem. 37:2679-2680.

Prokopy, R. J. 1972. Evidence for a marking pheromone deterring
repeated oviposition in apple maggot flies. Environ. Entomol.
1:326-332.

Price, P. W. 1970. Trail odors: recognition by insects parasitic
on cocoons. Science 170:546-547.

Price, P. W. 1972. Behaviour of the parasitoid *Pleolophus basizonus*
(Gravenhorst) (Hymenoptera: Ichneumonidae) in response to
changes in host and parasitoid density. Can. Entomol. 104:
129-140.

Price, P. W., and H. A. Tripp. 1972. Activity patterns of parasitoids on the Swaine jack pine sawfly, *Neodiprion swainei*
Middleton, and parasitoid impact on the host. Can. Entomol.
104:1003-1016.

Read, D. P., P. P. Feeny, and R. B. Root. 1970. Habitat selection
by the aphid parasite *Diaeretiella rapae* (Hymenoptera: Braconidae) and hyperparasite *Charips brassicae* (Hymenoptera: Cynipidae). Can. Entomol. 102:1567-1578.

Reuter, O. M. 1913. Lebensgewohnheiten und Instinkte den Insekten
bis zum Erwachen der Sozialaen Instinkte. Friedlander, Berlin.
448 p.

Rothschild, M. 1961. Observations and speculations concerning the
flea vector of myxomatosis in Britain. Entomol. Mon. Mag.
96:106-109.

Salt, G. 1937. The sense used by *Trichogramma* to distinguish
between parasitized and unparasitized hosts. Proc. R. Soc.
London. Ser. B. 122:57-75.

Southwood, T. R. E. 1961. The number of species of insect associated with various trees. J. Anim. Ecol. 30:1-8.

Spencer, K. A. 1969. The Agromyzidae of Canada and Alaska. Mem.
Entomol. Soc. Can. 64:1-311.

Townes, H. 1969. The genera of Ichneumonidae. Part 1. Mem.
Amer. Entomol. Inst. 11:1-300.

Vinson, S. B., and F. S. Guillot. 1972. Host marking: Source of
a substance that results in host discrimination in insect
parasitoids. Entomophaga 17:241-245.

Wheeler, W. M. 1910. Effects of parasitic and other kinds of castration upon insects. J. Exp. Zool. 8:377-438.

Zimmerman, E. C. 1958. Insects of Hawaii. Vol. 8. Lepidoptera: Pyraloidea. Univ. Hawaii Press, Honolulu. 456 p.

Zimmerman, E. C. 1960. Possible evidence of rapid evolution in Hawaiian moths. Evolution 14:137-138.

BIOCHEMICAL COEVOLUTION BETWEEN PARASITOIDS AND THEIR HOSTS

S. B. Vinson

Department of Entomology, Texas A&M University

College Station, Texas 77840, U.S.A.

The coevolution of a predator and its prey suggests that the prey is selected for predator avoidance and escape while the predator is selected for more efficient prey-finding and capture. The predator attacks and consumes its host and, thus being successful, would be expected to propagate. The parasite-host relationship is a similar reciprocating evolutionary relationship with one important difference--the host should not be killed. The parasite must not only locate the host but must constantly adapt to a developing and changing host, a situation resulting in a closely interwoven relationship. The continued development of a host in accord with the evolutionary development of the host species is of benefit to the parasite. It is probably advantageous for the parasite to evolve towards a condition where the parasite has a minimal affect on the host species. This would insure the continued existence of a viable host population.

The parasitoid-host relationship is akin to the predator- and parasite-host relationship, but with some unique aspects. The term parasitoid is used for arthropods where the egg and early larval stages are parasitic, but where the host is eventually killed. The adult is free living. In the parasitoid-host relationship, the parasitoid must not only locate the host, but must utilize the host for the progeny's benefit. Once the host has been parasitized, the parasitoid's progeny grow and develop at the expense of the host; ultimately killing the host. The host's future development is only of importance to the parasitoid. Thus the evolutionary strategy of the parasitoid is different from that of a parasite or predator. The parasitized host can no longer be considered as a member of the host species, but must be viewed as a container for the parasitoid. The evolutionary development of the parasitized host towards a

14

reduction in competition with the host species and towards an environment and habitat suitable for the parasitoid's development
would be advantageous. The evolution of the larval stage of the
parasitoid as expressed by the parasitized host, however, must
occur within the confines placed upon it by the host.

Although the biology of the parasitoid-host relationship has
been investigated for many years, only recently has the role played
by chemicals in this relationship been investigated. The identification and importance of chemicals in the behavior and physiology
of this relationship is beginning to yield new insight into the
uniqueness of the relationship.

Parasitic hymenoptera usually successfully parasitize only a
limited number of host species. There has been a great deal of
interest in determining the factors involved in successful parasitism and in the elucidation of the sign-stimuli (cues) which lead
to parasitism. Successful parasitism has been divided into a
sequence of steps. Doutt (1959), bringing together the observations
of Salt (1935) and Flanders (1953), divided the process necessary
for successful parasitism into 4 steps. These were: host-habitat-
location, host-finding, host-acceptance and host-suitability. I
propose a fifth step, that of host-regulation which I will describe
later.

In this discussion I point out the role played by chemicals
in the behavior and physiology of the parasitoid-host relationship.
No attempt has been made to give a complete account of the varied
and extensive literature covering the various aspects of the
parasitoid-host relationship, but rather to examine the relationship in light of chemical intermediaries with emphasis on several
parasitoids attacking members of the *Heliothis* complex (Lepidoptera:
Noctuidae).

HABITAT PREFERENCE

Female parasitoids often emerge or find themselves some distance
from their hosts and in an alien habitat. These females, in order
to be successful must locate their hosts. Although the chance
location of a host (Cushman 1926), or random searching (Rogers 1972),
which has been assumed in many parasitoid-host population models
(see Royama, 1971, for a review of such models), may suffice for
some species, such a situation would not appear to be advantageous
for parasitoids of a limited host range or host densities. The task
of host location may be reduced or confined as the female seeks out
a preferred habitat.

Only a limited amount of work has been carried out on habitat
preferences. After emergence, the female may seek a suitable

habitat with respect to physical factors such as temperature, humidity, light intensity and wind as well as food supply. Flanders (1937) reported that *Trichogramma evanescens* Westro. preferred a field habitat while *T. embryophagum* (Hartig.) was found most often in arboreal habitats and *T. semblidis* (Avr.) preferred a marsh habitat. He related these differences in habitat preference in part to their differences in crawling and flying habits. Simmonds (1954) found that *Spalangia drosophilae* Ashm., a parasitoid of *Drosophila*, was attracted to dampness and soil level within a grass habitat.

A similar indication of habitat preference can be given for *Cardiochiles nigriceps* Viereck, a parasitoid of *Heliothis virescens*, (F.), the tobacco budworm. As shown in Table 1, *C. nigriceps* are observed in open fields or pasture situations but are not seen in forested or heavily shaded areas. What factors are important in habitat preference are unknown, but we have observed that female *C. nigriceps* activity drops in the field when clouds obscure the sunlight. These observations suggest that light intensity may be one important factor. In the laboratory we have found that a high light intensity is necessary for optimal mating and oviposition (Vinson et al. 1973). Such differences in general habitat preference may limit the different habitats searched and hosts thus located.

HABITAT LOCATION

Although a female parasitoid may be located in a preferred habitat such as a forest or grassland, the host may only occur in specific situations within that habitat. It would be advantageous for a female to be able to find those habitats most likely to yield potential hosts. The cues important in host habitat location by the parasitoid could come from the host's food (plant), the host, stimuli resulting from the plant-host relationship or a combination of these stimuli (Fig. 1). Thorpe and Jones (1937), investigating olfactor conditioning of *Venturia* (*Nemeritis*) *canescens* (Grav.), reported that the parasitoid was first attracted to the host's food (oatmeal). Later, Thorpe and Caudle (1938) reported that mature adult *Pimpla ruficollis* Grav. were attracted to the odor of *Pinus* oil, and Arthur (1962), finding that parasitism by *Itoplectis conquisitor* (Say) was greater on Scots than Red pine, demonstrated that females were preferably attracted to the odor of Scots pine in an olfactometer.

We have observed *C. nigriceps* searching host free tobacco plants, grown in a glasshouse, after the plants were placed in an open area (Table 1). These results suggest that *C. nigriceps* is first attracted to the plant regardless of the presence or absence of hosts. A number of other studies demonstrating the

Table 1. Observations on female *Cardiochiles nigriceps* visiting tobacco plants in different locations, with or without *Heliothis virescens* larvae.

Plant Condition and Location	Number Females Searching Plant (1 hr.)	Number Females Landing on Plant
Full Sun, Host Free, Undamaged	12	2[c]
Full Sun, Host Free, Damaged[a]	8	14
Full Sun, Host Present, Damaged[b]	9	23
Forest Shade, Host Free, Undamaged	0	0
Forest Shade, Host Free, Damaged[a]	0	0
Forest Shade, Host Present, Damaged[b]	0	0

[a] Damaged by punching holes and scratching leaves

[b] Damage inflicted by *H. virescens* larvae

[c] Includes a female landing on a plant more than once

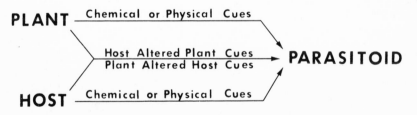

Fig. 1. Diagram showing the possible source of the first cues important in orientating a female parasitoid to its host.

role of the host's food (plant) in influencing or attracting the parasitoid have been carried out (Camors and Payne 1971, Nishida 1956, Read et al. 1970, Rice 1969, Streams et al. 1968 and Shahjahan 1974). The results of these studies demonstrate that many parasitoids do orient to factors from the host's food. Such

factors may serve to reduce or limit the area searched to those
areas most likely to contain hosts.

In other cases the role of the food source of the host in
orienting the parasitoid is not clear. *Nasonia* (*Mormoniella*)
vitripennis (Walker) is attracted to carrion-breeding dipterous
pupae contaminated with liver which had contained pupae but not
fresh or rotted liver or clean pupae (Edwards 1954, 1955, Wylie
1958). This contrasts with the results of Laing (1937), who reported that *N. vitripennis* was attracted to meat which had never
contained hosts. His results suggest that food liberated factors
are involved.

In some situations host liberated materials may be involved
in directing the parasite to the host habitat. Camors and Payne
(1971) reported that bark beetle pheromones attracted some of the
beetle's parasitoids and Sternlicht (1973) reported two parasitoids
responsive to pheromones of the California red scale. The utilization of a volatile compound liberated by a potential host as a cue
to host location by a parasitoid may be expected in situations
where either the stage releasing the compound is attacked or is
present along with the potential host stage. Both the food plant
or host derived attractants, are probably volatile factors perceived
by the female at some distance from the host. These "long range"
factors serve to bring the female into a potential host habitat.
In the case of host liberated volatiles, they also serve in host
detection.

HOST SEARCHING

Once the parasitoid has reached a potential host habitat she
must begin a search for the host. With *C. nigriceps* we have found
that once the host's food plant is located, the female initiates a
second plant-directed host oriented-searching behavior. During
this phase of host searching, the female is observed to scan the
food plant of the host (Fig. 2). This behavior is exemplified by
the female flying with outwardly directed antennae up and down the
plant stems and leaves 2 or 3 centimeters from the surface. This
pattern of movement is repeated from leaf to leaf and plant to
plant. During this phase of host searching the female is observed
occasionally to land on the plant and briefly palpate a small area
with her antennae. Close examination of areas visited by the
parasitoids revealed a damaged area of plant tissue.

In an effort to examine the role that damaged plant tissue
might play as a cue to the parasitoid, we placed glasshouse grown
tobacco plants free of insect damage, plants mechanically damaged
by punching small holes or scratching the leaf surface and plants
with host-produced damage, in a pasture. The pasture was located

Fig. 2. Female *Cardiochiles nigriceps* showing characteristic scanning of plant for potential hosts.

about 1/4 mile from cotton. Observations of the *C. nigriceps* examining these plants are given in Table 1. Parasitoids did not land on undamaged plants as frequently as mechanically damaged or host damaged plants. Further, the damage due to host feeding resulted in both an increased frequency of landing as well as increases in the length of time spent on the plant.

Females approaching small areas of damaged plant tissue usually landed briefly, touched the damaged tissue once or twice with their antennae and if the damage was not produced by a potential host, the female resumed scanning the plant. In some cases the female did not land, but appeared to touch the damaged area with her antennae. Plant tissue damaged by *H. virescens*, however, caused a change in this behavior. Females touching a *Heliothis*-damaged area would place both antennae on the plant around the area of damage and rub the surface. Instead of flying, the female will "excitedly" crawl over the plant surface rubbing the substrate with her antennae. These results suggest that the female can distinguish between host and non-host damaged plant tissue. As reported by Vinson (1968) host larvae secrete a chemical during feeding that the female appears to perceive resulting in host detection.

Once an area which has or has had a potential host is detected

by the parasitoid, the female must locate the host proper.   The
factors involved in this aspect of host location are either host
derived or intimately associated with the host and may be influenced
by non-host factors.

With *C. nigriceps* we have referred to this initial host-produced
material as a host-seeking stimulant (Vinson and Lewis 1965).   The
material from *H. virescens* which initiates host seeking by the parasitoid is located in the mandibular gland of the host (Vinson 1968).
Activity has also been observed in the frass and extracts of the
cuticle, although neither source is as active as the mandibular
gland.   When a female *C. nigriceps* contacts the host-seeking stimulant she becomes very excited and places the distal third of both
antennal flagella on the substrate (Fig. 3).   The female will pull
the antennae in contact with the substrate toward her, stroking
the surface and is able to follow a "trail" of the material.   This
activity often leads to the host.

There has been a great deal of work on host-seeking stimulants
or kairomones in recent years.   A kairomone is a chemical which is
produced by an organism that gives an adaptive advantage to the
receiving organism (Whittaker and Feeny 1971).   Kairomones probably
serve the producer in some way and the term is used here only from
the point of view of the receiver.   Corbet (1971) reported an epi-
deictic (dispersing) pheromone from the mandibular glands of
Anagasta kuehniella, (Zeller), the flower moth, which would elicit
ovipositional movements in *V. canescens*.   Other workers have observed activity in the frass of hosts (Hendry et al. 1973, Quednau
1967, Quednau and Hubsch 1964), although the sources have not been
determined.

In other parasitoid-host systems other sources of host-seeking
stimulants or compounds involved in host location occur.   Laing
(1937) and Lewis et al. (1971) reported a material from the scales
of adult moths that stimulated the egg parasitoid, *T. evanescens*.
Vinson (1975) has reported the presence of a chemical in the ovary
of the female moth that stimulates the egg-larval parasitoid,
Chelonus texanus Cresson.

As shown in Table 2 chemicals involved in host seeking have
been identified for only 4 species.   Mudd and Corbet (1973) have
reported that the mandibular gland of *A. kuehniella* contains a
single component with a molecular weight of about 392.   As pointed
out by Jones et al. (1971) and Vinson et al. (1975) these compounds
can be remarkably specific.   A change in the position of the
methyl branch one carbon unit on the molecule will destroy activity.
In addition the compounds are most active in the nanogram range and
are thus not as active as pheromones which often act in the picogram
range (Jacobson 1968, Beroza 1970 and Birch 1974).   There also is an
important dose effect.   As pointed out by Corbet (1973), response

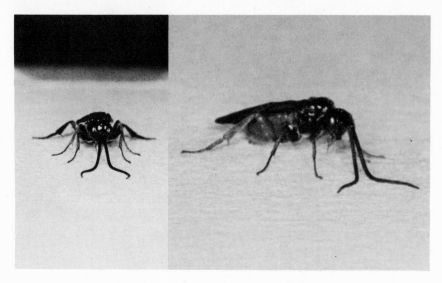

Fig. 3. Female *Cardiochiles nigriceps* showing characteristic antennal position in response to the mandibular secretions of its host.

by the parasitoid to extracts of the mandibular gland decreased on either side of an optimum concentration.

Another point that should be brought out is that several of the host-seeking stimulants thus far identified are rather large molecular weight materials and thus would be expected to have a low volatility. As shown by Vinson (1968), *C. nigriceps* does not appear to perceive the chemical or the presence of hosts from a distance but only upon contact with the host-seeking stimulant.

In several parasitoid-host relationships host odor has been reported to play a role in attracting or orientating the parasitoid (Bartlett and Lagace 1961, Bartlett 1953, Carton 1971, 1974, Spradberry 1970a, Weseloh 1974) although it is not yet established whether these odors play a major role as long range or short range factors. Hendry et al. (1973) has shown that heptanoic acid from the frass of the potato tuberworm, *Phthorimaea operculella*, will attract *Orgilus lepidus* while an unidentified contact chemical elicited ovipositor thrusting after the parasitoid antennally examined the frass. Their results suggest that the heptanoic acid acts as an intermediate in host habitat location and a contact chemical is actually involved in host-seeking. Upon closer examination the cues involved in host location may consist of a sequence of several chemicals each serving to reduce the area of search. Under such a concept volatile compounds (odors) from the

Table 2. Parasitoids for which the host-seeking stimulants (kairomones) have been identified.

Parasitoid	Hosts	Host Searching Stimulants	Ref.
Microplitis croceipes (Cresson)	*Heliothis virescens* (F.) *Heliothis zea* (Boddie)	13-Methyl Hentriacontane	Jones et al. 1971
Trichogramma evanescens Westro.	eggs of a wide range of hosts	Tricosane	Jones et al. 1973
Orgilus lepidus Muesebeck	*Phthorimaea operculella* (Zell.)	n-Heptanoic acid	Hendry et al. 1973
Cardiochiles nigriceps Viereck	*Heliothis virescens* (F.)	12-Methyl Hentriacontane 13-Methyl Dotriacontane 13-Methyl Tritriacontane	Vinson et al. 1975

plant and/or host would serve to bring the parasitoid into a host-containing area while a contact chemical, a chemical of low volatility or a chemical only perceived at a high concentration would identify those areas which once contained or presently contains potential hosts.

In many parasitoids host seeking may further be divided according to whether the antennae or the ovipositor is involved. The importance of the antennae in host seeking has been indicated by a number of authors (Spradbery 1969, 1970b, Greany and Oatman 1972, Richerson and Borden 1972, Richerson et al. 1972, Weseloh 1971). Hays and Vinson (1971) reported that the antennae were the primary organs involved in host seeking by *C. nigriceps* and that females lacking antennae were unable to locate a host or follow a trail of the host-seeking stimulant. Further examination of the antennae of the female *C. nigriceps* has revealed some unique "bent tip" receptors (Fig. 4) (Norton and Vinson 1974). These receptors have been correlated with host seeking activity (Table 3) suggesting their possible involvement in host location. However, as noted by Hays and Vinson (1971) a female would oviposit if the tarsi or ovipositor touched the host.

More recent observations with *C. nigriceps* demonstrate the importance of the ovipositor in host location. Although *H. virescens* larvae are sometimes exposed on a leaf, they often occur within the buds and seed pods of their host plants with the entrance covered with frass. When female *C. nigriceps* locates such an area she palpates the frass with her antennae and then begins stabbing or probing the area with her ovipositor. Ovipositor probing also occurs when a female encounters cracks and crevices in response to a synthetic host-seeking stimulant.

Another example of the importance of both the antennae and ovipositor in host searching is demonstrated by *C. texanus* where a chemical initiates antennal searching and the rough texture of the substrate appears to initiate ovipositor probing (Vinson 1975). Many other examples where the parasitoid, after contacting frass, undertakes drilling or ovipositor searching have been reported (Spradberry 1970b, Hendry et al. 1973, Fisher 1959, Quednau and Hubsch 1964). The data indicate that chemoreception with both the antennae and the ovipositor may be involved in host location. There is no reason to believe that both organs respond to the same chemicals or that the ovipositor is incapable of responding to chemical stimuli during drilling and ovipositor thrusting that aid in directing the ovipositor towards the host. The presence of chromoreceptors on the ovipositor has been shown for a number of parasitoids (see review of Fisher, 1971), although their importance has most often been in connection with their role in host acceptance and discrimination.

Fig. 4.   Scanning electron micrograph of the terminal antennal
segment of a female *Cardiochiles nigriceps* showing the unique
bent-tip receptors (arrow).

PARASITOID CONDITIONING

The role of chemicals in providing cues to host habitat and
host location and the effects of these chemicals on the behavior
as outlined above suggest that host specificity is regulated in
part by the presence or absence of a chain of chemical cues. Doutt
(1959) pointed out the importance of genetic preadaptation as an
essential factor in host specificity. There is, however, information to suggest that host specificity is not entirely due to a
sequence of parasitoid-host relationships but, after determining
host specificity on broad phylogenetic lines, that host specificity
is due to ecological factors (Cushman 1926). Townes (1962) pointed
out that the pattern of search does not by itself determine host
specificity. Taylor (1974) thus suggested that the proportion of
host specificity under genetic control must reside in host acceptance

Table 3.   The relationship between the behavioral response of
C. nigriceps to oral secretions of the host, the number of antennal
segments and the bent-tipped chemoreceptors present on those
segments.

Antennal Segments[1]	Bent-tipped[2] Chemoreceptors	Average[3] Response
Complete antenna	98.6	3.0
35 segments remaining	86.2	3.0
30	60.6	2.7
20	18.0	2.2
10	1.6	1.0
5	0.0	0
0	0.0	0
Scape and pedicel removed	0.0	0

[1] Number of antennal segments remaining after desegmentation

[2] Average number of bent-tipped chemoreceptors present on all
   remaining flagella segments

[3] Scored from zero to three with a score of three given for a
   strong positive response

and suitability.

The challenge to the supposition that host searching is under
genetic control is brought out by the well documented phenomenon
of learning in insects (Alloway 1972) and in the case of insect
parasitoids, the work of Thorpe (1938, 1943), Arthur (1966, 1971),
Taylor (1974) and Gross et al. (1975).   Arthur (1966), working
with a non-host specific parasitoid, *I. conquisitor*, reported that
the female learns to concentrate her search to productive habitats.
In a later paper (Arthur 1971) using a host-specific parasitoid,
V. canescens, showed that the female could be conditioned to
associate a "normal" odor with the presence of hosts.   Taylor (1974)
expanded on the above studies using *V. canescens* and showed that
the female can learn to hunt hosts in a novel environment and that

a model using the learning of two cues fits better than a model using only one. These results, however, do not reduce the importance of searching or the supposition that host searching and specificity at the habitat and host locating level are under genetic control.

This point can best be illustrated by examining the response of *C. nigriceps*, which is host specific (Chamberlin and Tennet 1926, Lewis and Vinson 1971), to other hosts and the parasitism of its habitual host on various plants on which the host occurs. We have collected many *C. nigriceps* from tobacco, spider flower, ground cherry (*Physalis* spp.) and Toadflax (*Linaria canadensis* (L.)). *C. nigriceps* has also been shown to be the most abundant parasite of *H. virescens* on Deergrass (*Rhexia mariana* L.), cotton and tobacco (Neunzig 1969) and Carolina cranesbill (*Geranium carolinianum* L.) (Snow et al. 1966). Shepard and Sterling (1972) reported that *C. nigriceps* is one of the more abundant parasitoids of *H. virescens* on cotton in Texas. On the other hand, J. W. Smith (Personal communication) has not observed *C. nigriceps* as a parasite of *H. virescens* on peanuts. *C. nigriceps* has been observed attacking *H. zea* (Boddie) on cotton (Lewis and Brazzel 1966), but is not observed searching peanuts or corn where *H. zea* more frequently occurs. This suggests that the parasitoid's host range is limited by factors that direct the parasitoid to the host's food plant. Once the food plant is located, further specificity is produced by a chain of cues that lead to the potential host. For example, *C. nigriceps* will accept for oviposition both *H. zea* and *H. virescens* which occur on cotton, but will not attack the cotton leaf perforator, *Bucculatrix thurberiella* (Busck), or the yellow-striped armyworm, *Spodoptera ornithogalli* (Guenec) which also occur on cotton. *C. nigriceps* will not attack the tobacco hornworm, *Manduca sexta* Johannson, which occurs along with *H. virescens* on tobacco. As shown in Table 4, *C. nigriceps* is not stimulated by secretions (frass or feeding damage) of the tobacco hornworm, cotton leaf perforator or yellow-striped armyworm and these larvae are not attacked when placed in a petri dish with female *C. nigriceps*.

Most of these same larvae are attacked and a few eggs deposited when such larvae are treated with the searching stimulant identified from *H. virescens* showing that such a material may serve as a factor determining host specificity (Table 5). As shown in Table 4, the beet armyworm, *Spodoptera exigua* (Hbn.) is readily attacked when exposed to *C. nigriceps* in the laboratory and eggs are deposited although the beet armyworm is not often observed in habitats searched by *C. nigriceps*.

Although host specificity may be determined in part by habitat selection and chemicals from the host that provide cues to their location, the host may occur in several different habitats or on different food sources. In addition, host populations may fluctuate

Table 4. Response of *Cardiochiles nigriceps* to a number of potential hosts.

Host	Common Name	Antennal[1] response	Ovipositional[2] activity	%[3] eggs deposited
Heliothis zea	Corn earworm	2.6	30(30)	81
Heliothis virescens	Tobacco budworm	3.8	30(30)	100
Heliothis phloxiphaga		3.7	30(30)	96
Heliothis subflexa		4.0	30(30)	100
Feltia subflexa	Granulated cutworm	1.8	4(20)	0
Spodoptera exigua	Beet armyworm	2.5	15(20)	68
Spodoptera frugiperda	Fall armyworm	1.2	5(20)	0
Spodoptera ornithogalli	Yellow striped armyworm	0.4	0(25)	0
Elasmophalpas lignosellus	Lesser cornstalk borer	3.2	1(20)[4]	0
Manduca sexta	Tobacco hornworm	1.3	2(15)	50
Trichoplusia ni	Cabbage looper	0.8	1(15)	0
Galeria mellonella	Wax moth	0.2	1(20)	0
Pectinophora gossypiella	Pink bollworm	0	0(25)	0
Anthonomus grandis	Boll weevil	1.3	9(20)	3
Musca domestica	House fly	0	0(15)	0
Bucculatrix thurberiella	Cotton leaf perforator	0.8	0(15)	0

[1] Average score with 4 given when the female places antennae in the substrate or host and rubs the substrate. 0 is no antennal response.

[2] The number of hosts accepted for stinging with the number exposed in parenthesis.

[3] The percentage of those hosts stung that revealed an egg upon dissection.

[4] The larva would quickly bend its body flipping out of the parasitoid's way. When a larva was touched the parasitoid could not attack successfully.

Table 5. Response of *Cardiochiles nigriceps* to "unattractive" hosts after treatment of host with a solution of the parasite-searching stimulant.

Host	Common name	Ovipositional[1] activity	% eggs deposited[2]
Spodoptera ornithogalli	Yellow-striped armyworm	3(20)	33
Feltia subterranta	Granulated cutworm	14(20)	3
Spodoptera frugiperda	Fall armyworm	18(20)	20
Anthonomus grandis	Boll weevil	18(20)	10
Galeria mellonella	Wax moth	16(20)	38
Musca domestica	House fly	11(20)	0

[1] The number of hosts accepted for stinging after application of the host-seeking stimulant. The number exposed is given in parenthesis.

[2] The percent hosts stung revealing an egg upon dissection.

from one food source to another at different times of the year as
exemplified by *Heliothis* populations (Lincoln 1972). It would not
only appear advantageous to the female parasitoid to locate and
search potential host habitats and host food sources, but to concentrate in habitats where success has been achieved. As shown by
Gross et al. (1975) *M. croceipes* can be conditioned to search particular host plants by first exposing them to host frass from larvae
feeding on the host plant in question. As pointed out by Gross,
exposure of *M. croceipes* to frass activates an inherent fixed action
of host-seeking directed to a specific set of cues. Attempts to
stabilize or condition the female by the use of the host-seeking
stimulant were unsuccessful (Gross et al. 1975). This does not
negate the possibility that compounds from the host, such as the
searching stimulant, are involved in the conditioning. The host-
seeking stimulant is found in frass (Lewis and Jones 1971) and
could initiate the conditioning while host food (plant) factors
in combination with the host-seeking stimulant, could give direction
to the conditioning. Such a multi-component system would lend
support to Taylor's two factor model (Taylor 1974). Parasitoid
conditioning within this context would allow for a degree of flexi-
bility within the confines of host habitat and host location
specificity dictated by genetic parameters.

HOST ACCEPTANCE

Once the female parasitoid has located a potential host the
next step is the acceptance of the host. Host acceptance, however,
can be subdivided into two different parameters. One deals with
factors from the host which are important in egg release (host
identification) and the other deals with factors that reduce the
acceptance of already parasitized hosts (host discrimination).

Host Identification

The factors involved in host identification and egg release are
difficult to separate from host location. There has been a great
deal of literature on host acceptance. Factors such as odor (Picard
1922, Madden 1968), movement (Arthur 1961, Richerson and DeLoach
1972), sound (Lloyd 1956), sight (Salt 1935, Laing 1938), shape
(Lingren et al. 1970, Eijsackers and Lenteren 1970, Weseloh 1970),
and electromagnetic radiation (Richerson and Borden 1971) have been
implicated as being important in host acceptance.

Host identification by *C. nigriceps* and *Campoletis sonorensis*
(Cameron) does not appear to be controlled by size, shape, movement
or the presence of a living host (Hays and Vinson 1971, Wilson et
al. 1974) although as shown by Wilson et al. (1974), these parameters
were minor factors that influence host acceptance. The presence of

a specific factor in the hemolymph did not appear necessary for
oviposition by *C. nigriceps*. When larvae were extracted with
chloroform-methanol they were not recognized as potential hosts.
If the host-seeking stimulant was applied to the cuticle, the
larvae were stung but no eggs were deposited (Hays and Vinson 1971).
These results show that host-seeking stimulant is a necessary component, but that other factors are necessary for egg release. This
is also borne out in Tables 4 and 5.

Some of the most interesting work on host acceptance has been
conducted by Arthur et al. (1969) with *I. conquisitor*. These
authors showed that a factor in the hemolymph of the host stimulated
I. conquisitor to lay eggs. Later the active factors were found to
be due to the presence of amino acids and sugars and that egg laying
could be elicited by serine, arginine, leucine and $MgCl_2$ (Arthur
et al. 1972, Hegdekar and Arthur 1973). Recently Rajendram and Hagen
(1974) reported that saline and amino acids would stimulate *Trichogramma* to oviposit into artificial eggs.

Host Discrimination

Salt (1934) reported that *T. evanescens* females left a factor
on their host that inhibited further attack. He further demonstrated
that the odor could be removed by washing. Salt (1937) also found
that *T. evanescens* could perceive that host eggs were parasitized
after drilling with the ovipositor. These inhibitory factors were
later termed "spoor factors" by Flanders (1951) although these compounds should be referred to as host marking pheromones or allomones
depending on whether they are intra or inter-specific.

Host discrimination can occur at several different points along
the chain of cues leading to the host, although host discrimination
is most often observed once the host is located. Both DeBach (1944)
and Price (1970) have reported that parasitoids avoid areas previously searched and Price (1970) reported inter-specific discrimination. In many cases, particularly with egg parasitoids, the host
is marked after one encounter (Hokyo et al. 1966; Rabb and Bradley
1970). After oviposition, the female may be observed to move over
the egg surface for a minute or more (Rabb and Bradley 1970), and
it is assumed that this post-ovipositional activity is involved in
marking.

Studies with several parasitoid species of *Heliothis* have shown
that *C. nigriceps* discriminates between nonparasitized and super-
parasitized hosts although it is less likely to discriminate against
a host parasitized only once. Similar results were obtained for
M. crociepes and *C. sonorensis* (Vinson 1972a, Vinson and Guillot
1972). With these parasitoids, host marking occurred but was not
active enough to prevent a degree of superparasitism. Such activity

may have survival value to the parasitoid as some ovipositional
attempts are aborted due to the aggressive behavior of the host.
Due to the aggressive host behavior, oviposition and host marking
are, by necessity, accomplished quickly. An aborted ovipositional
attempt could result in host marking yet leave the potential host
available for attack. Bakker et al. (1972) developed a model
describing egg distribution based on host discrimination. These
authors reported that a model where the chance of an egg being
deposited decreased as the number of eggs increased provided the
best fit.

Although host marking has been reported for a number of parasitoids, the source of the factors responsible for host discrimination has only been investigated in a few instances. Simmonds
(1954) suggested that odors from the hemolymph after ovipositor
puncturing of the cuticle or odors from a fluid left by the
ovipositor might be responsible for host discrimination. Lloyd
(1956) suggested a residual chemical from the accessory gland was
responsible. The source of the factors responsible for host discrimination by *C. nigriceps* was found to be Dufour's gland (Fig. 5)
(Vinson and Guillot 1972), and the active components were later
isolated in the hydrocarbon fraction of the gland (Guillot et al.
1974). Dufour's gland was also found to be responsible for host
discrimination by *M. croceipes* and *C. sonorensis* (Guillot and
Vinson 1972a, Vinson and Guillot 1972).

In *C. sonorensis* it was found that the external marking pheromone lost its effectiveness in less than 6 hours. Hosts previously
parasitized six hours or more were attractive and were stung but no
new eggs were found upon dissection. These results, similar to those
of Salt (1937) and Greany and Oatman (1971, 1972), show that females
could distinguish between parasitized and nonparasitized larvae upon
ovipositor insertion (Guillot and Vinson 1972a). Other cases of
host discrimination have been reported, many occurring in later
stages of parasitism and detected by the female following ovipositor
insertion (Fisher 1961, King and Rafai 1970, Griffiths 1971). Fisher
and Ganesalingam (1970) have shown that changes in the hemolymph
occur after parasitism and they suggest such changes may be responsible for discrimination. Guillot and Vinson (1972a) have shown
that the female injects a material from the ovary which is responsible for the internal host discrimination. Whether this factor
acts directly or indirectly is unknown.

HOST SUITABILITY

Once the host has been accepted it must be suitable for the
development of the parasitoid. The suitability of a host may
depend on several factors. The host may already have been parasitized resulting in competition for control of the host. Also,

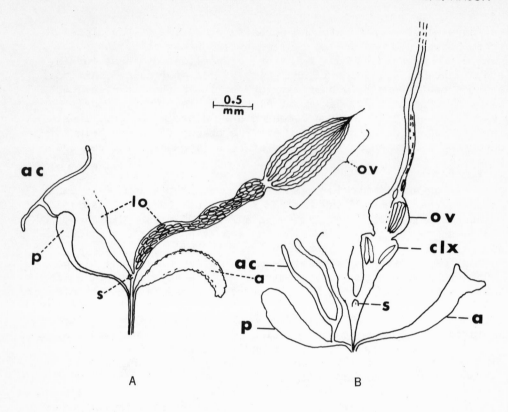

Fig. 5. A. Reproductive system from *Campoletis sonorensis*
(Ichneumonidae). B. Reproductive system from *Cardiochiles nigriceps* (Braconidae). ac = acid gland; a = Dufour's gland; clx =
calyx; lo = lateral oviduct; ov = ovarioles; p = poison sack; and
s = spermatheca.

the parasitoid must be able to overcome or evade the host's internal
defense mechanisms and the parasitoid must find the host nutritionally suitable for development.

    Most endophagous insect parasitoids require the entire host
for their development. In many cases hosts are parasitized more
than once resulting in multiparasitism (the parasitism by two or
more species) or superparasitism (two or more individuals of the
same species). In such situations the survival of the parasitoid
depends upon its ability to destroy its competitors. Successful
competition has been attributed to physical attack, physiological
suppression either by anoxia or secretion of a toxin, starvation
or changing of the host's conditions required for development.
The subject of host competition has been adequately reviewed by

Salt (1961) and Fisher (1961) and will not be examined further.

After oviposition the egg and the developing parasitoid must also evade or overcome the internal defense of the host. The subject of the defense reactions of insects to parasitoids and the resistance of parasitoids to the defense reactions of the host are beyond the scope of the present discussion. I refer the reader to the reviews of Salt (1963, 1968, 1970) and Whitcomb et al. (1974).

The nutritional suitability of a host has not been adequately investigated, although it is usually assumed that the host is in a nutritional form suitable for the developing parasitoid. Salt (1938) pointed out some of the problems in determining nutritional adequacy. The nutritional and accessory growth factor needs of most organisms are similar and parasitoids do not appear to be an exception (Yazgan 1972, Yazgan and House 1970). Thus, one would assume that most host insects should be nutritionally suitable. The problem of host nutritional suitability may not reside in nutritional inadequacy of the host but in the inability of the parasitoid to obtain the nutrition contained in the host's tissues.

The eggs and young larva must obtain all of their nutrition from the hemolymph during the early instars (Vinson and Barras 1970, Wilson et al. 1974, Corbet 1968). During this phase of the parasitoid's development, the parasitoid must also compete with the host's tissues for the limited nutritional supply in the hemolymph. As shown by Joiner and Vinson (unpublished data) (Table 6) there is an increase in lipids in the hemolymph at the expense of host body tissues after parasitism. Similar results were shown by Barras et al. (1969) for proteins and amino acids. The results would suggest that the ability of the parasitoid to alter the host's nutritional flow for its needs may be an important factor in host suitability.

HOST REGULATION

The effect of the parasitoid on the development, behavior, physiology and morphology of the host have been described by many authors (DeBach 1964, Sweetman 1963, Doutt 1963) but the importance of host regulation has not been appreciated. As pointed out by Fuhrer (1968) parasitism induces physical changes in the host and alters the hosts response to the environment. The parasitized hosts are ecologically different from nonparasitized hosts. Generally, the effect has been attributed to the parasitoid larva within the host as it begins to feed and develop. The role of the female parasitoid in regulating the host for her progeny's benefit has not been developed.

Table 6. The ratio of total and neutral lipids between nonparasitized *Heliothis virescens* larvae and larvae parasitized by
Cardiochiles nigriceps. Data in mg/mg wet wt for carcass
and fat body and mg/ml for hemolymph.

Tissue	Days after parasitism	Ratio of non-parasitized/parasitized	
		Total lipids	Neutral lipids
Carcass	5	0.96	0.84
	7	1.26	1.33
	9	1.59	1.76
Fatbody	5	1.51	1.62
	7	1.10	1.20
	9	0.73	0.82
Hemolymph	5	1.80	1.18
	7	0.68	0.84
	9	0.49	0.20

The importance of host regulation can best be described by
examining the relationship between *Heliothis* and several of its
parasitoids. As shown by Vinson and Barras (1970), Jones and
Lewis (1971), and Vinson (1972b), parasitized larvae do not develop
or grow as rapidly or as large as non-parasitized larvae. The
factor(s) responsible have been shown to be located in the calyx
or lateral oviduct (Fig. 5) of the parasitoids reproductive system
(Jones and Lewis 1971, Vinson 1972b, Guillot and Vinson 1972b)
although the poison gland was found to act synergistically for
C. nigriceps (Guillot and Vinson 1972b).

Solutions of the contents of the lateral oviduct from *C.
sonorensis*, when injected into 3rd instar larvae of *H. virescens*,
result in a cessation of feeding and development of treated larvae.
The effects are observed within one day after injection with such
larvae remaining in an arrested state as long as a month or until
they desiccate and die (Vinson 1972b). Although the mode of action
of the venom is unknown, it does not result in paralysis. Guillot
and Vinson (1973) showed that the venom from *C. nigriceps* did not
appear to effect the nervous system or musculature as parasitized
larvae were mobile and could be forced to chew and move food through
their digestive tract. The results suggest that the venom is
different from that of *Bracon* (Edwards and Sernka 1969) which blocks
the neuromuscular junction of its host (Piek and Engles 1969).

There is some suggestion that parasitism may interfere with the host's endocrine system. Johnson (1959) reported that the aphid parasite, *Aphidius plantensis* Brethes had a juvenilizing effect on the host. Vinson (1970) reported that teratocytes (cells which develop from the embryonic membrane of the parasite's egg) resulted in juvenilization of the host insect after their injection. Recently, Joiner et al. (1973) has shown that teratocytes from the host of *C. nigriceps* possesses juvenile hormone-like activity. Smilowitz (1974) and Iwantsch and Smilowitz (1975) have shown that the prothoracic glands of parasitized *Trichoplusia ni* larvae were smaller and that RNA synthesis was reduced in the glands of parasitized larvae compared to controls.

A number of authors have reported changes in various biochemicals within a host after parasitization. Changes in quantity and in some cases quality have been described for amino acids and proteins (Mellini and Callegarini 1967, Dahlman 1969, Barras et al. 1972, Barlow 1962, Smilowitz 1973, Junnikkala 1966, Leibenguth 1970), carbohydrates (Dahlman 1970, Dahlman and Herald 1971, Fuhrer 1972, Waller 1965) and lipids (Barras et al. 1970, Thompson and Barlow 1972, 1973). The role of venoms in initiating these biochemical changes has not been investigated.

The mode of action of venoms and their importance in regulating the host for the developing parasitoid may offer new insights into the parasitoid-host relationship. There is some evidence to suggest that host specificity may also depend on the effect of a venom on the host and the changes brought about by the venom may be necessary for the developing parasitoid. When the poison gland was surgically removed from female *C. nigriceps* which were then permitted to parasitize the host, the percentage of progeny that ultimately emerged was significantly reduced and those parasitoids that did emerge were retarded (Guillot and Vinson 1972b). It also has been reported that when developing larvae were transplanted to nonparasitized hosts, they failed to develop and were encapsulated (Vinson 1972c).

Host suitability was described by Salt (1938) to be due in large measure to unsuitable chemical constituents necessary for proper development of the parasite. Thus a host might be unsuitable due to improper nutrition or due to presence of deleterious chemicals. Although such a possibility may occur, the failure of the parasitoid to regulate the host's internal environment may also be an important factor influencing "host suitability". An example of this possibility can be given by examining the work of Dahlman and Vinson (1975). In this study the trehalose levels in *H. virescens* were compared to the levels obtained after parasitism by *C. nigriceps* and *M. croceipes*. The trehalose levels decreased in the hemolymph of larvae parasitized by *C. nigriceps* while they were elevated in larvae parasitized by *M. croceipes*. The elevation of trehalose in host larvae parasitized by *M. croceipes* was measurable one day after parasitism suggesting that the female parasitoid may have been responsible. Such an increase in carbohydrate reserves in the hemolymph, possibly at the expense of the body reserves, might be of major importance to a hemolymph feeder such as *M. croceipes* that must obtain all its nutrition from that source. An elevation of trehalose might have less adaptive significance to a parasitoid that consumes its host. The failure of a host to respond to such host regulating chemicals would render it unsuitable.

SUMMARY

The evolutionary development of the plant and the plant feeding host must not be overlooked when considering the evolutionary development of the parasitoid host relationship. Each step in the sequence of cues leading to host habitat and host location by the female parasitoid serves to reduce the area of search and increases the probability of successful parasitism. Such a series of cues reducing the area a female must search would also be expected to result in a high degree of efficiency in locating potential hosts and would reduce the types of habitats searched and hosts that can be located resulting in increased specificity.

The plant often provides the first cues in the chain of steps leading to the host. A number for plant feeding insects, such as *Heliothis*, have been shown to exhibit host-plant preferences (Jermy et al. 1968) that can lead to biotypes and host plant changes. Some hosts may be able to escape their parasitoids through a host plant preference change where the new host plant lacks the necessary cues to attract or orientate the parasitoid. Changes in the plant, brought about by natural selection or plant breeding, may not greatly reduce the plants attractiveness to the plant feeder, but may result in the loss of the factors necessary to orientate the parasitoid. Either of these situations coupled with the host plant preferences by both sexes could lead to the rapid development of a new species.

An analogous example is provided by *H. subflexa* which serves as a host for *C. nigriceps* in the laboratory (Lewis and Vinson 1971), although less than 5% of the field collected larvae are attacked (Lewis et al. 1967). *H. subflexa* has only been reported from ground cherry (*Physalis* spp.) and the plant is searched by *C. nigriceps*. However, *H. subflexa* larvae grow and develop within the enlarged calyx of the ground cherry and are generally not accessible to the parasitoid.

The demonstration that parasitoids can be conditioned to search certain habitats along with the demonstration of the importance of a series of chemical cues in habitat and host location suggest that the behavioral system is an open one. Although much of the behavioral pattern involved in host location is probably genetically controlled, the parasitoid is able to learn certain environmental cues that leads the female to habitats where success has previously been achieved.

Recent work has established that chemicals provide the female parasitoid with cues to the host's location and are involved in identification and acceptance of the host. However, these same chemicals must also serve the host in some way. The function of the compounds thus far identified as host-seeking stimulants have not been determined, although Corbet (1971) has shown that a host-seeking secretion from *A. kuehniella* acts as a dispersing pheromone and sex pheromones have been shown to attract certain parasitoids (Camors and Payne 1971, Sternlicht 1973).

The importance of host-seeking stimulants in host specificity is demonstrated in Table 5 where several species of otherwise unacceptable larvae are accepted for oviposition after the host seeking stimulant is applied. It is interesting to note that the addition of the host seeking stimulant to the yellow-striped army-worm did not significantly increase its attractiveness to *C. nigriceps*. This would suggest that those larvae either have a repellant or have duplicated the marking pheromone of *C. nigriceps*. The housefly and granulated cutworm offers yet another situation where the larvae are stung after being treated with the host-seeking stimulant, but are not accepted as a suitable site for oviposition.

After the host has been accepted for oviposition, the female appears to inject materials, along with the eggs. The specificity of these factors and their functions are for the most part unknown. The evolutionary significance of host regulation can only be speculated, although the importance of host regulation in providing a suitable internal environment for the successful development of the parasitoid's progeny and the alteration, selection and development of a suitable niche for the parasitoid's progeny are some possibilities.

As shown in Fig. 6 there are a number of steps leading to successful parasitism of the host by *C. nigriceps*. Chemicals appear to play an important role in many of these steps. Much more work will be necessary with *C. nigriceps* and its host, and many other parasitoid-host relationships before an understanding of the evolutionary development of this unique interaction can be unraveled. It does appear that chemicals do play a major role in the relationship.

ACKNOWLEDGEMENTS

Approved as publication TA 11553 by the director of the Texas Agricultural Experiment Station. Supported in part by Cotton, Inc.

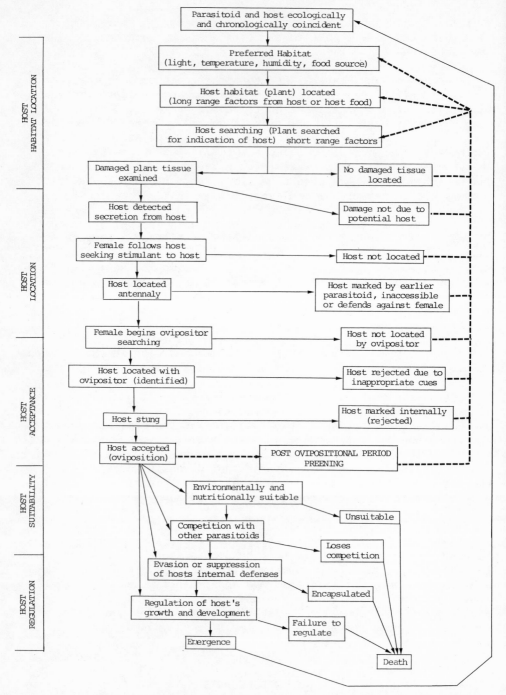

Fig. 6. Factors involved in successful parasitism by *Cardiochiles nigriceps*.

(CI-199) from funds made available through the USDA and NSF grant GB-24282.

LITERATURE CITED

Alloway, T. M. 1972. Learning and memory in insects. Ann. Rev. Entomol. 17:43-56.

Arthur, A. P. 1961. The cleptoparasitic habits and immature stages of *Eurytoma pini* Bugbee (Hymenoptera: Chalcidae), a parasite of the European shoot moth, *Rhyacionia buoliana* (Schiff.) (Lepidoptera: Olethreutidae). Can. Entomol. 93:655-660.

Arthur, A. P. 1962. Influence of host tree on abundance of *Itoplectis conquisitor* (Say) (Hymenoptera:Ichneumonidae), a polyphagous parasite of the European pine shoot moth *Rhyacionia buoliana* (Schiff) (Lepidoptera:Olethreutidae). Can. Entomol. 94:337-347.

Arthur, A. P. 1966. Associative learning in *Itoplectis conquisitor* (Say) (Hymenoptera:Ichneumonidae). Can. Entomol. 98:213-223.

Arthur, A. P. 1971. Associative learning by *Nemeritis canescens* (Hymenoptera Ichneumonidae). Can. Entomol. 103:1137-1141.

Arthur, A. P., B. M. Hegdekar, and W. W. Batsch. 1972. A chemically defined, synthetic medium that induces oviposition in the parasite *Itoplectis conquisitor* (Hymenoptera:Ichneumonidae). Can. Entomol. 104:1251-1258.

Arthur, A. P., B. M. Hegdekar, and C. Rollins. 1969. Component of the host hemolymph that induces oviposition in a parasitic insect. Nature (London) 223:966-967.

Bakker, K., H. J. P. Eijsackers, J. C. Van Lenteren, and E. Meelis. 1972. Some models describing the distribution of eggs of the parasite *Pseudeucoila bochei* (Hym., Cynip.) over its hosts, larvae of *Drosophila melanogaster*. Oecologia 10:29-58.

Barlow, J. S. 1962. An effect of parasitism on hemolymph electroherograms. J. Insect Path. 4:274-275.

Barras, D. J., R. L. Joiner, and S. B. Vinson. 1970. Neutral lipid composition of the tobacco budworm, *Heliothis virescens* (Fab), as effected by its habitual parasite, *Cardiochiles nigriceps* Viereck. Comp. Biochem. Physiol. 36:775-783.

Barras, D. J., R. L. Kisner, W. J. Lewis and R. L. Jones. 1972. Effects of the parasitoid, *Microplitis croceipes*, on the haemolymph proteins of the corn earworm, *Heliothis zea*. Comp. Biochem. Physiol. 43:941-949.

Barras, D. J., G. Wiygul and S. B. Vinson. 1969. Amino acids in the haemolymph of the tobacco budworm *Heliothis virescens* (F.) as effected by its habitual parasite. Comp. Biochem. Physiol. 31:707-714.

Bartlett, B. R. 1953. A tactile ovipositional stimulus to culture *Macrocentrus ancylivorus* on an unnatural host. J. Econ. Entomol. 46:525.

Bartlett, B. R., and C. E. Lagace. 1961. A new biological race
    of *Microterys flavus* introduced into California for the con-
    trol of lecaniine coccids, with an analysis of its behavior
    in host selection. Ann. Entomol. Soc. Amer. 54:222-227.
Beroza, M. ed. 1970. Chemicals controlling insect behavior.
    Academic Press, New York. 170 p.
Birch, M. C. ed. 1974. "Pheromones". American Elsevier. New York.
Camors, F. B., Jr., and T. L. Payne. 1971. Response of *Heydenia
    unica* (Hymenoptera: Pteromalidae) to *Dendroctonus frontalis*
    (Coleoptera: Scolytidae) pheromones and a host-tree terpene.
    Ann. Entomol. Soc. Amer. 65:31-33.
Carton, Y. 1971. Biologie de *Pimpla instigator* F. (Ichneumonidae,
    Pimplinae) I. Mode de percerption de l'hote. Entomophaga 16:
    285-296.
Carton, Y. 1974. Biologie de *Pimpla instigator* (Ichneumonidae:
    Pimplinae). III Analyse experimental de processus de re-
    connaissance de l'hote-chrysalidae. Ent. Exp. Appl. 17:265-278.
Chamberlin, F. S., and J. N. Tennet. 1926. *Cardiochiles nigriceps*
    Vier., an important parasite of the tobacco budworm, *Heliothis
    virescens* Fab. J. Agri. Res. 33:21-27.
Corbet, S. A. 1968. The influence of *Ephestia kuehniella* on the
    development of its parasite *Nemeritis canescens*. J. exp.
    Biol. 48:291-304.
Corbet, S. A. 1971. Mandibular gland secretions of larvae of the
    flour moth, *Anagasta kuehniella*, contains on epideictic
    pheromone and elicits oviposition movements in a hymenopteran
    parasite. Nature 232:481-484.
Corbet, S. A. 1973. Concentration effects and the response of
    Nemeritis canescens to a secretion of its host. J. Insect
    Physiol. 19:2119-2128.
Cushman, R. A. 1926. Location of individual hosts versus systematic
    relation of host species as a determining factor in parasitic
    attack. Proc. Ent. Soc. Wash. 28:5-6.
Dahlman, D. L. 1969. Haemolymph specific gravity, soluble total
    protein and total solids of plant-reared, normal, and para-
    sitized diet-reared tobacco hornworm larvae. J. Insect
    Physiol. 15:2075-2084.
Dahlman, D. L. 1970. Trehalose levels in parasitized and non-
    parasitized tobacco hornworm, *Manduca sexta*, larvae. Ann.
    Entomol. Soc. Amer. 63:615-617.
Dahlman, D. L., and F. Herald. 1971. Effects of the parasite,
    Apanteles congregatus, on respiration of tobacco hornworm,
    Manduca sexta larvae. Comp. Biochem. Physiol. 40:871-880.
Dahlman, D. L., and S. B. Vinson. 1975. Trehalose and glucose
    levels in the hemolymph of *Heliothis virescens* parasitized
    by *Microplitis croceipes* or *Cardiochiles nigriceps*. J.
    Comp. Physiol. Biochem. In press.
DeBach, P. 1944. Environmental contamination by an insect para-
    site and the effect on host selection. Ann. Entomol. Soc.
    Amer. 37:70-74.

DeBach, P.   1964.   Biological control of insect pests and weeds.
    Reinhold, New York.   844 p.
Doutt, R. L.   1959.   The biology of parasitic Hymenoptera.   Annu.
    Rev. Entomol. 4:161-182.
Doutt, R. L.   1963.   Pathologies caused by insect parasites.
    pp. 393-422.   In E. A. Steinhaus, ed.   Insect pathology, an
    advanced treatis.   Academic Press, New York.   Vol. 2.
Edwards, R. L.   1954.   The host-finding and oviposition behavior
    of *Mormoniella vitripennis* (Walker) (Hym., Pteromalidae), a
    parasite of muscoid flies.   Behaviour 7:88-112.
Edwards, R. L.   1955.   How the hymenopteran parasite *Mormoniella
    vitripennis* (Walker) finds its host.   Anim. Behav. 3:37-38.
Edwards, J. S., and T. J. Sernka.   1969.   On the action of *Bracon*
    (Hymenoptera: Braconidae).   Venom. Toxicon. 6:303-305.
Eijsackers, H. J. P., and J. C. van Lenteren.   1970.   Host choice
    and host discrimination in *Pseudeucoila bochei* (Hym., Cynip.).
    Neth. J. Zool. 20:414.
Fisher, R. C.   1959.   Life history and ecology of *Horogenes chryso-
    stictos* (Hymenoptera, Ichneumonidae), a parasite of *Ephestia
    sericarium* Scott (Lepidoptera, Phycitidae).   Can. J. Zool.
    37:429-446.
Fisher, R. C.   1961.   A study in insect multiparasitism.   II. The
    mechanism and control of competition for possession of the
    host.   J. Exp. Biol. 38:605-628.
Fisher, R. C.   1971.   Aspects of the physiology of endoparasitic
    Hymenoptera.   Biol. Rev. 46:243-278.
Fisher, R. C., and V. K. Ganesalingam.   1970.   Host hemolymph changes
    composition after insect parasitoid attacks.   Nature 227:191-192.
Flanders, S. E.   1937.   Habitat selection by *Trichogramma*.   Ann.
    Entomol. Soc. Amer. 30:208-210.
Flanders, S. E.   1951.   Mass culture of California red scale and
    its golden chalcid parasites.   Hilgardia 21:1-42.
Flanders, S. E.   1953.   Variations in susceptibility of citrus-
    infesting coccids to parasitization.   J. Econ. Entomol. 46:
    266-269.
Fuhrer, E.   1968.   Parasitaere Ueraenderungen am Wirt and ihre
    Oekologische Bedeutung Fued den Parasiten.   Entomophaga 13:
    241-250.
Fuhrer, E.   1972.   Abnorme Glykogenspeicherung in Larven von *Pieris
    brassicae* L. (Lep., Pieridae) als Folge des Parasitismus von
    Apanteles glomeratus L. (Hym., Braconidae).   Z. Angew. Entomol.
    70:370-374.
Greany, P. D., and E. R. Oatman.   1971.   Demonstration of host
    discrimination in the parasite *Orgilus lepidus* (Hymenoptera:
    Braconidae).   Ann. Entomol. Soc. Amer. 65:375-376.
Greany, P. D., and E. R. Oatman.   1972.   Analysis of host dis-
    crimination in the parasite *Orgilus lepidus* (Hymenoptera:
    Braconidae).   Ann. Entomol. Soc. Amer. 65:377-383.
Griffiths, K. J.   1971.   Discrimination between parasitized and
    unparasitized hosts by *Pleolophus basizonus* (Hymenoptera:

Ichneumonidae). Proc. Entomol. Soc. Ontario 102:83-91.

Gross, H. R., W. J. Lewis, and R. L. Jones. 1975. Kairomones and their use in Management of entomophagous insects. III. Stimulation of *Trichogramma achalae* and *Microplitis croceipes* with host seeking stimulants at time of release to improve their efficiency. Environ. Entomol. In press.

Guillot, F. S., and S. B. Vinson. 1972a. Sources of substances which elicit a behavioral response from insect parasitoid, *Campoletis perdistinctus*. Nature (London) 235:169-170.

Guillot, F. S., and S. B. Vinson. 1972b. The role of the calyx and poison gland of *Cardiochiles nigriceps* in the host-parasitoid relationship. J. Insect Physiol. 18:1315-1321.

Guillot, F. S., and S. B. Vinson. 1973. Effect of parasitism by *Cardiochiles nigriceps* on food consumption and utilization by *Heliothis virescens*. J. Insect Physiol. 19:2073-2082.

Guillot, F. S., R. L. Joiner, and S. B. Vinson. 1974. Host discrimination and isolation of hydrocarbons from Dufour's Gland of a braconid parasitoid. Ann. Entomol. Soc. Amer. 67:720-721.

Hays, D. B., and S. B. Vinson. 1971. Host acceptance by the parasite *Cardiochiles nigriceps* Viereck. Anim. Behav. 19: 344-352.

Hegdekar, B. M., and A. P. Arthur. 1973. Host hemolymph chemicals that induce oviposition in the parasite *Itoplectis conquisitor* (Hymenoptera:Ichneumonidae). Can. Entomol. 105:787-793.

Hendry, L. B., P. D. Greany, and R. J. Gill. 1973. Kairomone mediated host-finding behavior in the parasitic wasp *Orgilus lepidus*. Entomol. Exp. Appl. 16:471-477.

Hokyo, N., M. Shiga, and F. Nakasuji. 1966. The effect of intra- and interspecific conditioning of host eggs on the ovipositional behavior of two scelionid egg parasites of the southern green stink bug, *Nezara viridula* L. Jap. J. Ecol. 16:67-71.

Iwantsch, G., and Z. Smilowitz. 1975. Relationship between the parasitoid *Hyposoter exiguae* and *Trichoplusia ni*: Prevention of host pupation at the endocrine level. J. Insect Physiol. In press.

Jacobson, M. 1968. Insect sex attractants. Interscience Publishers, John Wiley and Sons, Inc. New York.

Jermy, T., F. E. Hanson, and V. G. Dethier. 1968. Induction of specific food preference in lepidopterous larvae. Entomol. Exp. Appl. 11:211-230.

Johnson, B. 1959. Effect of parasitisation by *Aphidius platensis* Brethes on the developmental physiology of its host, *Aphis craccivara* Koch. Entomol. Exp. Appl. 2:82-99.

Joiner, R. L., S. B. Vinson, and J. B. Benskin. 1973. Teratocytes: A source of juvenile hormone activity in a parasitoid-host relationship. Nature new Biol. 246:120.

Jones, R. L., and W. J. Lewis. 1971. Physiology of the host-parasite relationship between *Heliothis zea* and *Microplitis croceipes*. J. Insect Physiol. 17:921-927.

Jones, R. L., W. J. Lewis, M. Beroza, B. A. Bierl, and A. N. Sparks. 1973. Host-seeking (Kairomones) for the egg parasite *Tricho-gramma evanescens*. Environ. Entomol. 2:593-595.

Jones, R. L., W. J. Lewis, M. C. Bowman, M. Beroza, and B. A. Bierl. 1971. Host-seeking stimulants for parasite of corn earworm. Isolation, identification, and synthesis. Science 173:842-843.

Junnikkala, E. 1966. Effects of braconid parasitization on the nitrogen metabolism of *Pieres brassicae* L. Ann. Acad. Sci. Fennicae ser. A. 4, Biologica 100:1-83.

King, P. E., and J. Rafai. 1970. Host-discrimination in a gregarious parasitoid, *Nasonia vitripennis* (Walker) (Hym. Pteromalidae). J. Exp. Biol. 53:245-254.

Laing, J. 1937. Host-finding by insect parasites. I. Observation on the finding of hosts by *Alysia manducator*, *Mormoniella vitripennis* and *Trichogramma evanescens*. J. Anim. Ecol. 6: 298-317.

Laing, J. 1938. Host-finding by insect parasites. II. The chance of *Trichogramma evanescens* finding its hosts. J. Exp. Biol. 15:281-302.

Leibenguth, F. 1970. Verunderungen der Haemolymphe Ausgewachsener *Ephestia* Raupen nach Infection mit *Mattesia dispora*. Z. Parasitenkd. 33:235-245.

Lewis, W. J. 1970. Study of species and instars of larval *Heliothis* parasitized by *Microplitis croceipes*. J. Econ. Entomol. 63: 363-365.

Lewis, W. J., and J. R. Brazzel. 1966. Biological relationships between *Cardiochiles nigriceps* and the *Heliothis* complex. J. Econ. Entomol. 59:820-823.

Lewis, W. J., and R. L. Jones. 1971. Substance that stimulates host-seeking by *Microplitis croceipes* (Hymenoptera: Bracon-idae), a parasite of *Heliothis* species. Ann. Entomol. Amer. 64:471-473.

Lewis, W. J., and S. B. Vinson. 1971. Suitability of certain *Heliothis* (Lepidoptera: Noctuidae) as hosts for the parasite *Cardiochiles nigriceps*. Ann. Entomol. Soc. Amer. 64:970-972.

Lewis, W. J., J. R. Brazzel, and S. B. Vinson. 1967. *Heliothis subflexa*. A host for *Cardiochiles nigriceps*. J. Econ. Entomol. 60:615-616.

Lewis, W. J., A. N. Sparks, and L. M. Redlinger. 1971. Moth odor: A method of host finding by *Trichogramma evanescens*. J. Econ. Entomol. 64:557-558.

Lincoln, C. 1972. Seasonal abundance. p. 2-5. *In* Distribution, abundance and control of *Heliothis* species in cotton and other host plants. Southern Cooperative Series Bull. 169.

Lingren, P. D., R. J. Guerra, J. W. Nickelsen, and C. White. 1970. Hosts and host age preference of *Campoletis perdistinctus*. J. Econ. Entomol. 63:518-522.

Lloyd, D. C. 1956. Studies of parasite oviposition behavior I. *Mastrus carpocapsae* Cushman (Hymenoptera: Ichneumonidae). Can. Entomol. 88:80-89.

Madden, J. 1968. Behavioral responses of parasites to symbiotic
    fungus associated with *Sirex noctilio* F. Nature (London)
    218:189-190.

Mellini, E., and C. Callegarini. 1967. Ricerce elettroforetiche
    sulle proteine dell'emolinfa delle larve di *Anagasta kuehniella*
    (Lep. Pyralidae) parasitizzate de *Devorgilla canescens*.
    (Grav.) (Hym. Ichneumonidae). Boll. Ist. Ent. Univ. Bologna
    28:241-252.

Mudd, A., and S. A. Corbet. 1973. Mandibular gland secretion of
    larvae of the stored products pests *Anagasta kuehniella*,
    Ephestia cautella, *Plodia interpunctella* and *Ephestia elutella*.
    Entomol. Exp. Appl. 16:291-293.

Neunzig, H. H. 1969. The biology of the tobacco budworm and the
    corn earworm in North Carolina. N. C. Agric. Exp. Stn. Tech.
    Bull. 196, 76 p.

Nishida, T. 1956. An experimental study of the ovipositional
    behavior of *Opius fletcheri* Silvestre (Hymenoptera: Braconidae)
    a parasite of the melon fly. Proc. Hawaiian Entomol. Soc.
    16:126-134.

Norton, W. N., and S. B. Vinson. 1974. A comparative ultrastruc-
    tural and behavioral study of the antennal sensory sensilla
    of the parasitoid, *Cardiochiles nigriceps*. J. Morphology
    142:329-350.

Picard, F. 1922. Parasites de *Pieris brassicae* L. Bull. Biol.
    Fr. Belg. 56:54.

Piek, T., and E. Engels. 1969. Action of the venom of *Microbracon
    hebetor* Say on larvae and adults of *Philosamia cynthia* Hubn.
    Comp. Biochem. Physiol. 28:603-618.

Price, P. W. 1970. Trail odors: recognition by insects parasitic
    on cocoons. Science 170:546-547.

Quednau, F. W. 1967. Notes on mating behavior and oviposition of
    Chrysocharis laricinellae (Hymenoptera: Eulophidae), a
    parasite of the larch casebearer (*Coleophora laricella*).
    Can. Entomol. 99:326-331.

Quednau, F. W., and H. M. Hubsch. 1964. Factors influencing the
    host-finding and host-acceptance pattern in some *Aphytis*
    species (Hymenoptera: Aphelinidae). S. Afr. J. Agric. Sci.
    7:543-554.

Rabb, R. L., and J. R. Bradley. 1970. Marking host eggs by
    Telenomus sphingis. Ann. Entomol. Soc. Amer. 63:1053-1056.

Rajendram, G. F., and K. S. Hagen. 1974. *Trichogramma* oviposition
    into artificial substrates. Environ. Entomol. 3:399-401.

Read, D. P., P. P. Feeny, and R. B. Root. 1970. Habitat selection
    by the aphid parasite *Diaeretiella rapae* (Hymenoptera:
    Cynipidae). Can. Entomol. 102:1567-1578.

Rice, R. E. 1969. Response of some predators and parasites of
    Ips confusus (LeC) (Coleoptera, Scolytidae) to olfactory
    attractants. Contrib. Boyce Thompson Inst. 24:189-194.

Richerson, J. V., and J. H. Borden. 1971. Sound and vibration
    are not obligatory host finding stimuli for the bark beetle

parasite, *Coeloides brunneri* (Hymenoptera: Braconidae). Entomophaga 16:95-99.

Richerson, J. V., and J. H. Borden. 1972. Host finding behavior and *Coeloides brunneri* (Hymenoptera:Braconidae). Can. Entomol. 104:1235-1250.

Richerson, J. V., and C. J. DeLoach. 1972. Some aspects of host selection by *Perilitus coccinellae*. Ann. Entomol. Soc. Amer. 65:834-839.

Richerson, J. V., J. H. Borden, and J. Hollingdale. 1972. Morphology of a unique sensillum placodeum on the antennae of *Coeloides brunneri* (Hymenoptera: Braconidae). Can. J. Zool. 50:909-913.

Rogers, D. 1972. Random search and insect population models. J. Anim. Ecol. 41:369-383.

Royama, T. 1971. A comparative study of models for predator and parasitism. Res. Popul. Ecol. (Kyoto) Suppl. 1:1-91.

Salt, G. 1934. Experimental studies in insect parasitism II. Superparasitism. Proc. Roy. Soc. Lond. B. 114:455-476.

Salt, G. 1935. Experimental studies in insect parasitism. III. Host selection. Proc. Roy. Soc. Lond. B. 117:413-435.

Salt, G. 1937. The sense used by *Trichogramma* to distinguish between parasitized and unparasitized hosts. Proc. Roy. Soc. Lond. B. 122:57-75.

Salt, G. 1938. Experimental studies in insect parasitism. IV. Host suitability. Bull. Entomol. Res. 29:223-246.

Salt, G. 1961. Competition among insect parasitoids. Symp. Soc. Exp. Biol. 15:96-119.

Salt, G. 1963. The defense reaction of insects to metazoan parasites. Parasitology 53:527-642.

Salt, G. 1968. The resistance of insect parasitoids to the defense reactions of their hosts. Biol. Res. 43:200-232.

Salt, G. 1970. The cellular defense reactions of insects. Cambridge Univ. Press. London. 118 p.

Shahjahan, M. 1974. *Erigeron* flowers as a food and attractive odor source for *Peristenus pseudopallipes*, a braconid parasitoid of the tarnished plant bug. Environ. Entomol. 3:69-72.

Shepard, M., and W. Sterling. 1972. Incidence of parasitism of *Heliothis* spp. (Lepidoptera: Noctuidae) in some cotton fields of Texas. Ann. Entomol. Soc. Amer. 65:759-760.

Simmonds, F. J. 1954. Host finding and selection by *Spalangia drosophilae* Ashm. Bull. Entomol. Res. 45:527-537.

Smilowitz, Z. 1973. Electrophoretic patterns in hemolymph protein of cabbage looper during development of the parasitoid *Hyposoter exiguae*. Ann. Entomol. Soc. Amer. 66:93-99.

Smilowitz, Z. 1974. Relationships between the parasitoid *Hyposoter exiguae* (Viereck) and cabbage looper, *Trichoplusia ni* (Hubner): Evidence for endocrine involvement in successful parasitism. Ann. Entomol. Soc. Amer. 67:317-320.

Snow, J. W., J. J. Hamm, and J. R. Brazzel. 1966. *Geranium carolinianum* as an early host for *Heliothis zea* and *H. virescens*

(Lepidoptera: Noctuidae) in The Southeastern United States with notes on associated parasites. Ann. Entomol. Soc. Amer. 59:506-509.

Spradbery, J. P. 1969. The biology of *Pseudorhyssa sternata* Merrill (Hym: Ichneumonidae) a cleptoparasite of siricid woodwasps (Hym). Bull. Entomol. Res. 59:291-297.

Spradbery, J. P. 1970a. Host finding by *Rhyssa persuasoria* (L.) an ichneumonid parasite of sircid woodwasps. Anim. Behav. 18:103-114.

Spradbery, J. P. 1970b. The biology of *Ibalia drewseni* Borries (Hymenoptera: Ibaliidae) a parasite of siricid woodwasps. Proc. R. Entomol. Soc. Lond. 45:104-113.

Sternlicht, M. 1973. Parasitic wasps attracted by the sex phero-mone of their coccid host. Entomophaga 18:339-342.

Streams, F. A., M. Shahjahan, and H. G. LeMasurier. 1968. Influ-ence of plants on the parasitization of the tarnished plant bug by *Leiophron pallipes*. J. Econ. Entomol. 61:996-999.

Sweetman, H. L. 1963. Principles of biological control. Wm. C. Brown, Co., Dubuque, Iowa. 560 p.

Taylor, R. L. 1974. Role of learning in insect parasitism. Ecol. Monogr. 44:89-104.

Thompson, S. N., and J. S. Barlow. 1972. Synthesis of fatty acids by the parasite *Exeristes comstockii* (Hymenoptera) and two hosts, *Galleria mellonella* (Lep.) and *Lucilia sericata* (Dip.). Can. J. Zool. 50:1105-1116.

Thompson, S. N., and J. S. Barlow. 1973. The inconsistent phos-pholipid fatty acid composition in an insect parasitoid, *Itoplectis conquisitor* (Say) (Lepidoptera: Pyralidae). Comp. Biochem. Physiol. 44:59-64.

Thorpe, W. H. 1938. Further experiments on olfactory conditioning in a parasitic insect. Proc. Roy. Soc. B. 126:370-397.

Thorpe, W. H. 1943. Types of learning in insects and other arthropods. Br. J. Psychol. 33:220-234.

Thorpe, W. H., and H. B. Caudle. 1938. A study of the olfactory responses of insect parasites to the food plant of their host. Parasitology 30:523-528.

Thorpe, W. H., and F. G. W. Jones. 1937. Olfactory conditioning in a parasitic insect and its relation to the problem of host selection. Proc. Roy. Soc. B. 124:56-81.

Townes, H. 1962. Host selection patterns in some nearctic ichneumonids (Hymenoptera). 11th Int. Congr. Entomol., Vienna 2:738-741.

Vinson, S. B. 1968. Source of a substance in *Heliothis virescens* that elicits a searching response in its habitual parasite, *Cardiochiles nigriceps*. Ann. Entomol. Soc. Amer. 61:8-10.

Vinson, S. B. 1970. Development and possible function of terato-cytes in the host-parasitoid association. J. Invert. Path. 16:93-101.

Vinson, S. B. 1972a. Competition and host discrimination between two species of tobacco budworm parasitoids. Ann. Entomol.

Soc. Amer. 65:229-236.

Vinson, S. B. 1972b. Effect of the parasitoid, *Campoletis sonorensis* on the growth of its host, *Heliothis virescens*. J. Insect Physiol. 18:1501-1516.

Vinson, S. B. 1973c. Factors involved in successful attack on *Heliothis virescens* by the parasitoid *Cardiochiles nigriceps*. J. Invertebr. Pathol. 20:118-123.

Vinson, S. B. 1975. Source of material in the tobacco budworm involved in host recognition by the egg-larval parasitoid *Chelonus texana*. Ann. Entomol. Soc. Amer. In press.

Vinson, S. B., and D. J. Barras. 1970. Effects of the parasitoid, *Cardiochiles nigriceps*, on the growth, development and tissues of *Heliothis virescens*. J. Insect Physiol. 16:1329-1338.

Vinson, S. B., and F. S. Guillot. 1972. Host marking: source of a substance that results in host discrimination in insect parasitoids. Entomophaga 17:241-245.

Vinson, S. B., and W. J. Lewis. 1965. A method of host selection by *Cardiochiles nigriceps*. J. Econ. Entomol. 58:869-871.

Vinson, S. B., F. S. Guillot, and D. B. Hays. 1973. Rearing of *Cardiochiles nigriceps* in the laboratory with *Heliothis virescens* as hosts. Ann. Entomol. Soc. Amer. 66:1170-1172.

Vinson, S. B., R. L. Jones, P. Sonnet, B. A. Bierl, and M. Beroza. 1975. Isolation, identification and synthesis of host seeking stimulants for *Cardiochiles nigriceps*, a parasitoid of the tobacco budworm. Entomol. Exp. Appl. In press.

Waller, J. B. 1965. The effect of the venom of *Bracon hebetor* on the respiration of the wax moth, *Galleria mellonella*. J. Insect Physiol. 11:1595-1599.

Weseloh, R. M. 1970. Influence of host deprivation and physical host characteristics on host selection behavior of the hyperparasite *Cheiloneurus noxius* (Hymenoptera: Encyrtidae). Ann. Entomol. Soc. Amer. 64:580-586.

Weseloh, R. M. 1971. Sense organs of the hyperparasite *Cheiloneurus noxius* (Hymenoptera: Encyrtidae) important in host selection processes. Ann. Entomol. Soc. Amer. 65:41-46.

Weseloh, R. M. 1974. Host recognition by the Gypsy moth larval parasitoid, *Apanteles melanoscelus*. Ann. Entomol. Soc. Amer. 67:585-587.

Whitcomb, R. F., M. Shapiro, and R. R. Granados. 1974. Insect defense mechanisms against microorganisms and parasitoids. *In* M. Rockstein (ed.) The physiology of insects. 2nd ed. Academic Press.

Whittaker, R. H., and P. P. Feeny. 1971. Allelochemics: Chemical interactions between species. Science 171:757-770.

Wilson, D. D., R. L. Ridgway, and S. B. Vinson. 1974. Host acceptance and ovipositional behavior of the parasitoid *Campoletis sonorensis* (Hymenoptera: Ichneumonidae). Ann. Entomol. Soc. Amer. 67:271-274.

Wylie, H. G. 1958. Factors that affect host finding by *Nasonia vitripennis* (Walk.) (Hymenoptera: Pteromalidae). Can. Entomol.

90:597-608.

Yazgan, S. 1972. A chemically defined synthetic diet and larval
    nutritional requirements of the endoparasitoid *Itoplectis
    conquisitor* (Hymenoptera). J. Insect Physiol. 18:2065-2076.

Yazgan, S., and H. L. House. 1970. An hymenopterous insect, the
    parasitoid *Itoplectis conquisitor*, reared axenically on a
    chemically defined synthetic diet. Can. Entomol. 102:1304-1306.

MODELS FOR PARASITE POPULATIONS

Rodger Mitchell

Zoology Department, Ohio State University

Columbus, Ohio 43210, U.S.A.

Ecological concepts and ecologists have played a prominent role in the development of two rather different kinds of models for dealing with consumer-resource relations. One kind can be called a population model because it represents a device employed to identify the ways in which a population is related to its resources and deals with some set of summary statistics that describe the population. The most fashionable population model is built around the concepts of r-species and K-species. The second kind of model is represented by Darwinian models. Darwinian models deal with relations within a single population and they have gained new utility through the use of energy budget concepts and life table statistics. Darwinian models must be species specific in order to deal with the interactions between the individuals of a population and be able to predict how the interactions may affect the contribution females make to the next generation.

Population models are useful in classifying populations according to the way in which they use resources but further use of the models is clearly limited. Darwinian models cannot be used to order or classify populations because they deal with the combinations of species traits that are peculiar to a particular species. Instead, Darwinian models are devices for explaining the dynamics of the process of natural selection when a suite of characters is being considered.

In the excitement generated by the recent marriage of ecology and evolution--and I suppose MacArthur and Wilson's "The Theory of Island Biogeography" (1967) might be the formal wedding of the two fields--enthusiasms have run far ahead of theory and population

models have been used in many different ways, some of which are not
appropriate. Studies dealing with r-species and K-species form a
welter of papers that demonstrate how data may or may not conform
to a set of postulates. Correspondences are often presented as
if they represented rigorous tests of falsifiable hypotheses.
Consequently there is a growing feeling of confusion about the
legitimate limits to the use of r and K (cf. Southwood et al. 1974
with Wilbur et al. 1974), and considerable emotion is associated
with the use of this population model (Van Valen and Pitelka 1974).
In order to explore the advantages and disadvantages of population
models with less controversy, I will look at the properties of an
unfamiliar population model developed some years ago (Mitchell
1964, 1967). Following that, some aspects of Darwinian models
will be considered in order to examine the values of alternative
approaches.

POPULATION MODELS

When ecologists become involved with evolution, they speak of
strategies and tactics and those terms appropriately identify the
unique contribution of ecologists. It is rather easy to define
the general strategies of host exploitation by parasites in terms
of their effect on the host life table. The parasitic Hymenoptera
considered at length elsewhere in this volume generally kill their
host and the effect on the host life table can be measured as an
alteration of the life table statistic for net reproductive rate
(R_O). This statistic is the product of age specific survivorship
and fecundity. If hosts are killed before reaching the reproductive age, then the parasite affects only survivorship and can be
called an l-parasite in conformity with life table notation.

The intercept of a parasite and its host can be described
with three statistics: First, there is incidence--the proportion
of hosts attacked. Second, there is the load of parasites which
is measurable either as the number of parasites over the entire
population of hosts or else as the load on parasitized hosts.
Finally the variance associated with the last two mean values must
be known. These statistics characterize the way a parasite
exploits its host and the limits for these statistics can be
derived from models for the extreme patterns of host exploitation.

The effect of l-parasites on their host is:

$$R_O = \Sigma(l-l_p)_x m_x$$

in which l_p is the frequency of parasitized hosts. The capacity
of a host population to adjust to various levels of mortality may
be testable in the laboratory and in some field situations.

Experimentation and mathematical models are likely to provide independent estimates for the limits to the value of l_p in well known systems.

If a parasite were to systematically attack and kill certain host phenotypes more often than others, the host population would evolve toward the phenotype that is less often attacked. Inter-actions involving such co-evolution of parasites and hosts will tend to drive populations of parasites into one of two patterns of host exploitation: Either all hosts will be equally susceptible to attack and the parasite will attack a small number of hosts throughout the host population, or else the parasite may be extremely variable and unpredictable in its attacks. In the latter case, the variance in parasite loads will always be very high and in the first case the variance will increase as the loads of parasites on a host increase.

The average load or density of parasites on parasitized hosts may vary but the maximum load limit for l-parasites will be determined by the relation of the needs of a parasite relative to the resources provided by a host item. When more than one para-site can mature on a host, then the efficiency with which a host population is exploited will be determined by the clumping of parasites. In this situation, efficiency is defined as maximizing the yield of parasites per host item exploited. If C is the capacity of the host and P the mean number of parasites per host, then the efficiency will be proportional to P/C which will range in value from $1/C$ when all hosts carry only one parasite to 1.0 when all hosts bear as many parasites as can mature on the host. When $P/C = 1.0$ there will be the largest possible yield of parasites for a given incidence of parasites.

The optimal strategy will quite clearly be a population in which $P=C$ on all parasitized hosts and, because the optimal situation is to have all parasitized hosts carry a full load, the statistics for the loads on parasitized hosts will be:

$$\bar{x}_p = C \qquad\qquad\qquad s_p^2 = 0$$

When the limit for changes in R_O is known, then the maximum value for l_p times the value of C will define the maximum stable popu-lation of parasites:

$$(l_p)_{max} \, C = \text{maximum parasite population}$$

Thus, in a direct and clearly operational fashion, it is possible to define the limits for an l-parasite and fully characterize the strategy that would maximize the parasite population. This is a basis for defining the perfect l-parasite and if the limit for l_p

and C are given, then a perfect l-strategy can be exactly defined.
Naturally the perfect l-species does not exist and the concept has
utility only in so far as it defines the way in which a suite of
characters are functionally related and provides a standard for
measuring and relating the characteristics of real populations.

For the most part mathematical population models define
extreme limits that do not exist in nature but are easily handled
mathematically while the real populations are all far from the
extremes and very difficult to define mathematically.

The parasites that affect the other function of R_O, fecundity,
are at the opposite extreme and these are the m-parasites, which
have an effect on the host indicated by:

$$R_O = \Sigma l_x (m-m_p)_x$$

In this system the maximum incidence can be 1.0 when the effect of
parasites on the host is small.

Unlike l-parasites in which the limit to the number of
parasites per host was a simple function of the host as a food
particle, m-parasites have limits fixed by the extent to which
parasites alter the host fecundity and the ability of hosts to
compensate for such losses. It is possible that the density
limits can be specified for a parasite through experimentation and
they will reflect the relation:

$$\text{parasite density} \sim \text{host } m_x$$

From this it follows that the maximum parasite population will be
produced when all hosts are parasitized and each carries the
maximum sustainable load of parasites. Such a system will have
zero variance for all measures of the numbers of parasites per
host. The derivation of statistics describing the pattern of
exploitation is simpler for m-parasites but any effective measure
of m-affects in field populations may be difficult.

The patterns of host exploitation at the two extremes can be
tabulated as to the functions involved and the determinants of the
limits.

l-parasites	Function	m-parasites
$(l-l_p) \ll 1.0$	Incidence	$l_p = 1.0$
host mass $\sim C$	Density on hosts	$C \sim (m-m_p)_x$
$s^2 \gg \bar{x}$	Variance of mean loads	$s^2 = 0$
$s_p^2 = 0$	Variance between loads	$s_p^2 = 0$

It is obvious that only one function of this tabulation is determined independently by a property of the parasite and that is the value of C for l-parasites. Everything else is a measure of the response of the host to parasitism and the ability of the host population to sustain the load of parasites. No parasite will have the traits defined by the extremes and so the system simply provides an orderly way of classifying and defining the parasite strategies in a way that allows them to be ordered along a continuum with l-parasites at one extreme and m-parasites at the other. The suites of traits associated with the different strategies are all mathematically defined and directly derived from an analysis of the host life table.

The life table model does not merely identify a pattern and then associate postulates and inferences in order to characterize strategies, as must be done in the case of r-species and K-species. Clearly an m-parasite fits rather closely the concept of a K-species but if the concept "K" is employed, there is no independent way to define or measure K-ness. On the other hand m-ness is a clearly definable suite of characteristics and the life table model above provides an exact statistical basis for measuring how closely a species approaches perfect m-ness.

When l-parasites are considered, it is clear that they are not necessarily r-species. The unpredictable localized l-parasites might be assumed to be r-species as a rule, while the widespread low-density l-parasites would be more likely to fit the concept of a K-species.

Regardless of whether l and m or r and K are employed, it is clear that both population models deal with the relations of a consumer to its resource and may identify the way a consumer might exploit its resource. Such schemes have three useful functions: 1) They allow a comparison of species independent of the biological peculiarities of the species, 2) They identify the relations between a suite of adaptive characters, and 3) They define the limits for stable consumer-resource relations and all consumer systems must lie within these limits.

EXAMPLES OF m-PARASITES

It is easiest to find examples of m-parasites because these parasites have a high incidence and a very precise parasite-host relation, hence, these will be used to illustrate the values and the problems of employing population models. There are several examples of mites that regulate their loads on a host. These include a mite parasitic on fresh-water mussels (Mitchell 1965), the moth ear mite (Treat 1957, 1958a,b) and a quill mite living on sparrows (Kethley 1971).

All of these parasites show an extremely tightly regulated
load and in every case the mites seem to have no measurable effect
on the host. Indeed, they lie far below the limit of the host to
sustain parasites.

1. The moth ear mite, *Myrmonyssus phalaenodectes* Treat (Treat
1957, 1958a,b). This mite completes its cycle within the ear of
a noctuid moth. These ears are large cavities in which the families
of several invading female mites can mature. The ears of moths are
essential for detecting the sonar of bats, and the moths respond
to bat sonar with a random fluttering flight to the ground (Roeder
and Treat 1957). One ear is adequate to hear bats and, as long as
moth ear mites are limited to one of the ears of a moth, they and
the moth will not suffer from predation by bats.

The major aspect of regulation is with respect to the site of
parasitism. When a female discovers a host, she always goes
through a very precise exploratory ritual. Should the host have
no mite on it, there will be a random selection of an ear and the
female will mark a trail from the mid-dorsal line to the ear she
enters. Later arrivals discover this trail during their ritual
exploration and follow the trail to the infected ear.

Careful observations by Treat (1958a) strongly suggest that
there is a precise pattern in the use of the resources of the ear.
Females enter the tympanic recess where they deposit eggs in a
clump and the egg clump is protected and defended from other
females. Feeding is done through the walls of the air sac, young
mites mate and rest in the counter-tympanic cavity and the cleanli-
ness of the colony is maintained because all the mites move far
away from the above sites to defecate.

The spatial limits and the regulation of behavior obviously
controls the population but there seem to be no controls as to the
numbers of invading females per host. Each ear can evidently
support at least three female mites and the incidence averages less
than 0.5. At these incidences very few (less than 0.2 per cent) of
the hosts would be expected to receive more than three mites when
encounters are random. The rarity of overloading probably means
that adaptations for numerical regulation are unnecessary.

Obviously the moth ear mite exhibits a very precise and well
regulated pattern of host exploitation but due to the low incidence,
the capacity of an ear may never be reached. Incidence is, of
course, a product of the abundance of dispersants times the probability of encountering a host. The details of this relation will
be explored below.

2. *Unionicola fossulata* Wolcott (Mitchell 1955, 1965). This

parasite of a fresh water mussel is limited to two females per
host and this is the absolute limit seen in extensive field
samples. The mean load of parasites is 1.96±0.08 mites per host
and standard error indicates the total variation over an annual
generation. Hence, both the variance in numbers of parasites for
the total population and the variance of loads on parasitized
hosts is significantly less than the mean. Regulation is even more
obvious when hosts are examined because the adult mites are limited
to one site on the host, the groove at the base of the gills. Each
female invades either the left or right side of the host and remains
there throughout her life. All parasitized hosts carry only one
or two female mites. This appears to conform to the pattern of an
m-parasite in which there is no apparent effect on the host,
variances are low and incidences are high (incidence = 0.8)

The hosts in some localities regularly support a parasitic
fauna of five species (Mitchell 1955) and in some localities only
U. fossulata is present (Mitchell 1965). The regulation of the
numbers of *U. fossulata* is independent of the presence or absence
of other species. This makes it obvious that the *m*-ness of this
species is not related to host capacity. The factor limiting the
numbers of females is the oviposition site. After females of
U. fossulata enter a mussel in August they begin to slowly produce
eggs that are precisely oriented as they are inserted in the margin
of the outer gill. About 15 eggs are produced a month and these
eggs accumulate over a ten month period prior to the almost simultaneous hatching which occurs in July. Unlike most foreign bodies,
the eggs do not provoke host responses. A possible reason for the
lack of host response is that if the tissues of the feeding groove
and those near the feeding groove did respond, it would disrupt
host feeding. Hence, it appears that the host responses are avoided
by either exploiting critical tissues that are unresponsive or by
having a precise oviposition pattern or both.

The eggs from one female will completely fill the outer gill
on her side. Further oviposition, or a second female, would over-
crowd the site, destroy the orientation of eggs and presumably
reduce survivorship of eggs. The population limit for *U. fossulata*
is, therefore, a limit defined by the capacity of safe oviposition
sites.

3.   The quill mite, *Syringophiloidus minor* (Berlese) (Kethley
1971). This mite lives inside the cavities of the flight and covert
feathers of house sparrows. It achieves incidences of 1.0 and is
regulated so that single invasions are 40 per cent more frequent
than expected under a random invasion. The variances are higher
than expected only because regulation occasionally breaks down.
Very large overloads are found in about 5 per cent of the feathers.

In this case the mite loads are not related to the parasite
load a host can support. It appears that the mites have evolved
into being quill dwellers in order to avoid the very effective
abilities of birds to control ectoparasites (Kethley and Johnston
1975). In exchange for the security of the quill, the mites are
limited by the space of the quill and, as Kethley (1971) argues,
intraspecific competition for space has resulted in tactics that
regulate the numbers of invading females, the life cycle pattern,
and the growth budget so as to maximize the yield of daughters
from each invading female.

Given the habits of the quill mite, it appears to be a pure
m-parasite but according to the population model it falls short
of being an m-parasite because it has not breached host defenses
and approached the capacities of the host to support a parasite
load.

Evaluation of m-Parasites

A quick look at the general biology of these three parasites
suggests that they are very close to having the mathematical properties of m-parasites. The correspondence is easily seen by
tabulating the expected and observed relations (Table 1). It is
quite clear that these species are at the extreme of having no
apparent effect on the host and this excludes any possibility of
there being a mix of l and m traits.

The three species can be ordered as to incidence from the
moth ear mite to the quill mite with 100 per cent incidence. The
reason for the low incidence in the moth ear mite seems to be the
problem of dispersal from moth to moth. At such low incidences
it is apparent that reductions in variance would not significantly
alter the efficiency or fitness of the mite. The other two mites
do have high incidences, and adaptive responses effectively
regulate the numbers of parasites per host in ways that conform
to the statistical expectations of the model. *Unionicola fossulata*
with an incidence of 0.8 and a density of 1.10 females per host
could presumably increase its total population to 2N (where N is
the number of hosts) rather than 1.1 x .8N so one might say that
the parasite population lies at about 40 per cent of its maximum
limit. The quill mite, on the other hand is at the limit. Each
parasitized feather is fully exploited and the yield per female
maximized.

But the limits for the parasite loads of each of the three
species (Table 1) are well below the theoretical limits defined
by the life table model. In every case the limits reflect the
biological peculiarities of the consumer-host relation not the

Table 1. The relations of the biology of three supposedly m-parasites to the limits defined by the population model for i and m strategies.

Function (expected limit)	*Myrmonyssus phalaenodectes*	*Unionicola fossulata*	*Syringophiloidus minor*
Effect on Host ($\sim m - m_p$)	None Measurable	None Measurable	None Measurable
Incidence ($i = 1.0$)	$i < 0.5$	$i = 0.8$	$i = 1.0$
Mean Load ($\bar{X} = C \sim$ host m_p)	$C \sim$ host discovery	$C \sim$ oviposition sites	$C = \dfrac{\text{feather vol.}}{\text{mite vol.}}$
Variance of Mean Load ($S^2 << \bar{X}$)	$S^2 \lesssim \bar{X}$	$S^2 << \bar{X}$	$S^2 << \bar{X}$
Variance in Parasite Loads ($S_p^2 << \bar{X}_p$)	$S^2 \lesssim \bar{X}_p$	$S_p^2 << \bar{X}_p$	$S_p^2 << \bar{X}_p$

capacity of the host to support parasites.

Conformity with the statistical expectations developed under the model is obtained whenever there is a biological limit to parasite densities. The life table model gives the maximum limit and Table 1 shows that a different suite of critical interactions come into play long before the maximum limit is reached. Obviously conformity to a majority of the expectations under an m-strategy does not necessarily mean that anything close to that strategy is being employed. In these examples none of the points of conformity with the m-strategy are explicable from the general model. They can be explained only if the biology of each system is studied. General theory fails to identify the factors operating in the origin and evolution of these m-parasites because the model for an m-species is the special case of the extreme at which host responses to mortality or a reduction of fecundity determine the density limits for parasites. Real species lying on the continuum between the extremes are limited by completely different sets of factors that determine an intermediate adaptive equilibrium before the limits to m-effects are reached. The end points of populations models cannot be used to define the direction of evolution and adaptive equilibria of species with mixed strategies and densities below the absolute limits.

For these reasons it is completely wrong to speak of l-selection and m-selection except for populations proven to be operating at the limits. These conclusions apply to other population models and emphasize the need for caution when speculating about adaptations and evolution from general population models.

DARWINIAN MODELS

If the alternatives of individual activity are not examined, we can speak of adaptations only in the sense that fins and shells are adaptations. Evolutionary models accounting for the origin, differentiation and maintenance of an adaptation require detailed information on the differences within populations. All too often Darwinian explanations become associated with the general population phenomena of regulation and reproductive potential. In order to avoid this source of confusion it is necessary to clearly define Darwinian fitness and the ways of measuring fitness. Except for populations moving toward extinction, a population with densities below average will tend to increase, and populations with above average densities will tend to decrease. There are limits to the oscillations in density. If R_O remains less than 1.0 over many cycles of resource abundance, a population will become extinct. Conversely no population with R_O consistently

over 1.0 can be stable. These facts are often used to argue that
an $R_O = 1.0$ is an adaptation, but to term it an adaptation is to
imply the possibility of alternative states, such as choosing
between legs, flippers or wings. There are no viable alternatives
to an average R_O of 1.0; this is a fixed algebraic property of all
sets of cohorts that persist in time. Biologists are often so
discomfited at letting survival be a chance phenomenon that they
unwittingly embrace models that require extant populations to have
some mechanism for evolving what is needed for the survival of the
population. If this cosmic argument is accepted as a definition
of a necessary evolutionary goal, it presumably requires a kind of
selection at a population level that can override simple Darwinian
selection for a greater production of young by each female.

Darwinian selection operates within cohorts to pass on the
genes of a female in proportion to her relative success in contributing individuals to the next generation. This is often presumed
to favor the traits of high fecundity because fitness (w) is
defined as the yield of young (n) from a female. Relative fitness
is measured as the deviation of the n of a female from the mean
of the females in her cohort. The fitness of females from different
cohorts cannot be compared unless it is clearly established that
the frequency of genotypes and the selective regimen of the different cohorts is comparable.

Any selection that results in maintaining the average value
of R_O in the face of Darwinian selection for above average R_O
must operate to subvert the differential transfer defined by $w = n$
but there is, as yet, no direct evidence or even a model showing
how this sort of selection might operate. Arguments concerning
the adequacy of Darwinian explanations for the observed patterns
require that Darwinian evolution be exhaustively explored with
rigor and precision and be shown to be lacking. One crucial
issue in this exploration is the matter of relating fitness to
the reproductive yield of a female.

In the simplest cases, fitness may be the number of viable
eggs released by a female, but the only cases that are likely to
be that simple are animals such as oysters, salmon and mayflies.
These animals release a store of eggs almost simultaneously in
space and time so that the Darwinian fitness of such females may
be the number of hatchable eggs. The numbers of hatchable eggs is
a measure of fitness, providing that all the newly hatched larvae
have the same probability of survival. The latter is obviously
difficult to prove even for animals that never care for their young.

This difficulty can be operationally defined by considering
the probability of young surviving to the adult stage and completing reproduction. The special cases mentioned above are all

instances in which the survival of larvae is likely to be independent
of the fecundity and behavior of the mother.

If the survivorship of young is related in any way to the
division of the pool of nutrients among the ova of a female or
else depends on some behavioral activity that either takes energy
from that available for ova or reduces the survivorship of the
female, then Darwinian fitness will involve complex tradeoffs
between fecundity, oviposition patterns and survivorship. The
true measure of fitness of a female is then nS (when S is the
survivorship of a young) and the effect of alternative patterns of
activity on S or n can be measured and developed with Darwinian
models.

Statistics of parasite loading were derived from population
models and it is necessary to show how these patterns might
increase fitness through the alteration of S under different
strategies. An explanation for the evolutionary process must
define the advantages and disadvantages of a given pattern. Each
of the end points of the population model required the variance
of the loads of parasitized hosts to be zero and the mean load to
equal the capacity. The benefits are defined by the population
model but the model does not define the cost of the behavioral
devices that generate parasite distributions of maximal fitness
and efficiency in host exploitation.

The best way to begin to measure costs is by means of a
single model that relates the elements responsible for generating
the distribution of mites on hosts so that both the effort and the
cost are specified in ways that are measurable. The Poisson distribution is an appropriate way to describe the behavior of an
undifferentiated parasite that randomly searches for a randomly
dispersed host and attaches to the first host it discovers. When
the distribution of parasites follows a Poisson distribution, the
variance will equal the mean, the incidence will be $1-e^{-\bar{x}}$ and the
mean for parasites per host will also be the mean number of
contacts per host. If all features of this system are held
constant except for host discovery, then some aspects of host
discovery can be modeled.

The parasite distribution will shift from a random pattern
toward a uniform distribution if already exploited hosts are
rejected. A parasite that rejects a host must find an alternate
host and in finding another host, the parasite will have to expend
more effort and time, as well as increase its exposure to predation.
These are the costs of obtaining a uniform distribution.

In such a system the mean can represent the mean number of
contacts and the incidence will then be predictable as

$1-e^{-\bar{c}}$ (in which $\bar{c}$ is the mean number of contacts of parasites with hosts). With data on incidence the theoretical value for $\bar{c}$ can be easily determined. The estimate for the increment of work that is required to search out added hosts and achieve a uniform distribution can be obtained as follows: If a random system has a mean load of 0.5 parasites per host, only 30 per cent of the hosts will be discovered and 9.1 per cent of the hosts will carry more than one parasite under a Poisson distribution. If fitness is maximal when there is only one parasite per host then the parasite must reject already parasitized hosts until 0.5 of the hosts have been discovered. When $1-e^{-\bar{c}}$ equals 0.5 the value for $\bar{c}$ will be 0.7, hence, the population of parasites must increase the frequency of contacts with hosts to 0.7 contacts per host, in order to find 0.5 of the hosts and achieve a uniform distribution. The added effort required to go from 0.5 to 0.7 contacts per host is 40 per cent. This use of the Poisson distribution and the biological values for c estimate the relative selective pressures in a form that can be rigorously tested.

This can be easily applied to the data on *U. fossulata*. The load is limited to two females per host and maximal fitness of females is achieved when there are one or two females per host. Consequently the load-related fitness of females will be 1.0 until a mean density of two mites per host is reached. Above that density fitness declines as a function of $2/\bar{x}$. The curve for maximum fitness (w_{max}) is discontinuous (Fig. 1) and converges with the curve for fitness under the Poisson at high densities.

At low mean values, the difference $w-w_{max}$ is very small and the frequency of undiscovered hosts very large, hence, a very small number of additional host contacts will produce a uniform distribution. The contact curve (Fig. 1) indicates the proportional increase in contacts needed to achieve w_{max}. Once the limits of two parasites per host is reached the relative gain in fitness (Δw) for going from a random to a uniform distribution declines (Fig. 1).

At the observed density of 1.1 female mites per host a mite with random host discovery and no rejection of hosts would loose 13 per cent in fitness because that fraction of the mites would be in excess of two female mites per host and these might either die or interfere with established females. In the latter case the loss of fitness would be even greater. If the incidence is increased to produce a more uniform distribution, then there must be more than one contact per host. The observed incidence of 0.8 can be taken as the performance of the mite in terms of host discovery and from Fig. 1 or solving for $\bar{c}$ when $1-e^{-\bar{c}} = 0.8$ it can be determined that 1.6 contacts per host are necessary in order

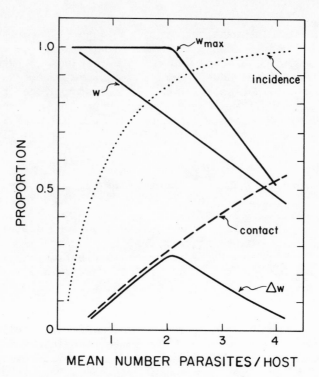

Fig. 1.   Expansion of the Poisson to define the elements of fitness
when the host capacity is two parasites.   The fitness (w) and
incidence under the Poisson are plotted. w_{max} is the fitness
achieved under a uniform distribution and the potential gain in
fitness (Δw) is the difference between the two fitness curves.
The contact curve indicates the proportional increase in host
contacts needed to attain w_{max}.

to achieve an incidence of 0.8.   To go from the expected incidence
of 0.67 to an incidence of 0.8 requires the numbers of host
contacts to be increased from 1.1 to 1.6, a 46 per cent increase
in the searching out of hosts.   This added effort results in a 13
per cent increase in fitness.

A model for evolution such as this has three obvious uses.
First, it indicates a baseline measure of fitness for an imaginary
mite with no special adaptations and, therefore, provides a basis
for measuring how much the various adaptations add to fitness.
The second way in which the model can be used is to contrast the
relative significance of different aspects of host exploitation.
There is no way at present of knowing how much the effort of
searching out additional hosts may cost a mite either in terms of
energy or in terms of survivorship, but this model directs

attention to the way the units of effort affect the units of
survivorship. Such an algorhythm shows the way in which field data
dealing with these phenomena could be used and may stimulate field
research.

The third use comes in evaluating strategies. A mite could
pre-empt resources in two ways. It was assumed above that the
later arrivals did not compete with established mites. Presumably
a phenotype that competed for sites and usually won would be ·
favored and competitive phenotypes would quickly dominate the
population. In this particular system, a competitive mite would
encounter fully occupied hosts 10 per cent of the time and this
would be its increment in fitness over non-competitive mites. But
once the non-competitive phenotype is eliminated, the population
must either continue to evolve new competitive traits or evolve
traits favoring either the resident or the attacker. The way in
which the equilibrium is established may influence the course of
evolution. Whichever of the equilibria are attained, the fact is
that any non-competitive mite that searches out alternative hosts
will have a 70 per cent chance of finding that the next host is
unoccupied. The cost relation between conflict and searching is:

Cost of winning x .10 : Cost of host discovery x .70

Of course the balance given here is for a mean load of 1.1 mites
per host. As loads increase, the frequency of unoccupied hosts
will decrease and at some point competition will represent an
alternative more likely to succeed than the discovery of new hosts.

With the very limited data available it is not possible to
explore the evolutionary alternatives open to *U. fossulata*. The
alternatives are the complex outcome of distributions that define
host availability, and mite abundances and habits. These are
sufficiently complex that general rules or relations cannot be
defined. The expansion of the Poisson does provide a framework
for measuring the consequences of various patterns of adaptation
and, for this reason, opens the prospect of being able to link
field studies together in new and instructive ways. Hence, it is
a model for testing field data and resolving the relative values
of adaptations that has great power in the analysis of field
populations but is not a basis for predictions.

In moth ear mites the incidence of mites regularly reaches
50 per cent but rarely exceeds that value by much. The capacity
of one moth ear has not been determined but it seems to be in
excess of three females, hence, random dispersion will not result
in very many overloadings. Although the elements of host discovery
are not known, this appears to be a system in which the prospects
for host discovery are low. At incidences of 0.5, only one-half

per cent of the hosts will have four or more mites. The system
appears to be one in which host discovery is so low that overloading is rare, therefore, it is adaptive for the mites to
remain on the host they contact and compete for sites on that host
rather than disperse to new hosts. Regulation in this case limits
mites to one site but does not impose absolute limits to the numbers
per site.

CONCLUSIONS

Models at the population level, such as the $r-K$ and $l-m$
continuua discussed here, define suites of adaptations that fit
together to form an integrated strategy of host exploitation. If
the adaptations can be directly measured, a population model will
provide the basis for an immensely useful biological classification
of species according to their strategy of host exploitation.

Since the models define the optimal strategy at the limits,
it is not appropriate to use population models as if they explained
the relations of real populations because real populations are not
at the end points and the complex balance of selection pressures
in intermediate situations admits to a variety of patterns of
optimization. In addition, population models deal only with
summary statistics that are not necessarily associated with the
determinants of differential fitness within populations. Hence,
the population models define strategies but they do not specify
the fitness coefficients that determine the path of Darwinian
evolution.

Models for natural selection must deal with differences of
fitness within a population and show how these fit together to
determine the overall fitness of individuals. Certain relations
concerning the discovery of hosts and parasite loading can be
defined through an extension of the Poisson. This will provide
the basis for Darwinian models but the form of the model must be
altered to fit the unique biology of the host-parasite relation.
It is impossible to predict whether generalities in the Darwinian
models will be found or not until more field systems are studied
in ways that will develop and test a series of specific Darwinian
models.

ACKNOWLEDGEMENTS

The final preparation of this manuscript was completed at the
Field Museum in Chicago where I began work in biology 35 years ago.
During that time the warm friendships and intellectual stimulation
given me by the staff of the museum has done much to mold my

interests. There is no way to single out individuals for I am indebted to them all for the opportunities that were opened to me at the Field Museum.

LITERATURE CITED

Kethley, J. B. 1971. Population regulation in quill mites (Acarina: Syringophilidae). Ecology 52:1113-1118.

Kethley, J. B., and D. E. Johnston. 1975. Resource tracking patterns in bird and mammal ectoparasites. Misc. Publ. Entomol. Soc. Amer. In press.

MacArthur, R. H., and E. O. Wilson. 1967. The theory of island biogeography. Princeton Univ. Press, Princeton, N.J. 203 pp.

Mitchell, R. 1955. Anatomy, life history and evolution of the mites parasitizing fresh-water mussels. Misc. Publ., Mus. Zool., Univ. Mich. 89. 28 pp.

Mitchell, R. 1964. A study of sympatry in the water mite genus *Arrenurus* (Family Arrenuridae). Ecology 45:546-558.

Mitchell, R. 1965. Population regulation of water mite parasitic on unionid mussels. J. Parasitol. 51:990-996.

Mitchell, R. 1967. Host exploitation of two closely related water mites. Evolution 21:59-75.

Roeder, K. D., and A. E. Treat. 1957. Ultrasonic reception by the tympanic organs of noctuid moths. J. Exp. Zool. 134:127-158.

Southwood, T. R. E., R. M. May, M. P. Hassell and G. R. Conway. 1974. Ecological strategies and population parameters. Amer. Natur. 108:791-804.

Treat, A. 1957. Unilaterality in infestations of the moth ear mite. J. N. Y. Entomol. Soc. 45:41-50.

Treat, A. 1958a. Social organization in the moth ear mite (*Myrmonyssus phalaenodectes*). Proc. 10th Int. Congr. Entomol. 2:475-480.

Treat, A. 1958b. A five year census of the moth ear mite in Tyringham, Massachusetts. Ecology 39:629-634.

Van Valen, L., and F. A. Pitelka. 1974. Commentary - intellectual censorship in ecology. Ecology 55:925-926.

Wilbur, H. M., D. W. Tinkle and J. P. Collins. 1974. Environmental certainty, trophic level, and resource availability in life history evolution. Amer. Natur. 108:805-817.

COURTSHIP IN PARASITIC WASPS

Robert W. Matthews

Department of Entomology, University of Georgia

Athens, Georgia 30602, U.S.A.

Courtship may be defined as any behaviors between conspecific individuals of opposite sex which facilitate mating. Such behaviors, with their incredible variety and obvious importance, have provided biologists with a vast and fascinating field of study. The evolution of such systems has, in particular, received considerable attention in recent years. For the vertebrates, this area has been capably reviewed by Orians (1969) and Trivers (1972).

For the Arthropods, an overview of the evolution of mating systems is provided by Alexander (1964). In addition, studies of courtship behavior have been undertaken in several insect groups, perhaps the most detailed of which are those on *Drosophila* (see Spieth 1974). Yet invertebrate courtship as a whole is still quite imperfectly known, as the parasitic Hymenoptera make abundantly clear.

The parasitic Hymenoptera belong to four superfamilies--the Ichneumonoidea, Chalcidoidea, Proctotrupoidea and Cynipoidea--and include a vast number of small to large insects, the majority of which live at the expense of their phytophagous or carnivorous relatives. In sheer numbers of kinds--i.e., species richness-- the parasitic wasps have few rivals. The Ichneumonidae alone constitute one of the largest families of insects, with an estimated 60,000 species placed in 25 subfamilies (Townes 1969), a diversity exceeding that of the well-studied vertebrate class Aves nearly ten-fold. The other major ichneumonoid family, the Braconidae, probably includes close to 25,000 species placed in 18 to 20 subfamilies (Matthews 1974). The classification of Chalcidoidea is not fully agreed upon, but approximately 25 families are recognized, with perhaps as many as 250,000 species in the world. The smallest

superfamily, the Cynipoidea, still includes at least 176 genera
and probably over 1400 species (Weld 1952). Best estimates in all
groups of parasitic wasps are that only from 10 to 25% of species
have even been described. Of course, far fewer have received any
serious study.

Parasitic wasps have an almost incredible range of life histories (see Clausen 1940, Doutt 1959, 1964, Askew 1971). The spectrum of hosts attacked is broad, ranging from ticks and spiders to
include virtually all orders of insects and all stages of hosts.
Yet in one sense they are rather uniform--all adults are free-
living, obtaining nourishment from nectar, honeydew, host feeding,
etc. Although a few groups (Cynipidae, Agaonidae, Torymidae,
Eurytomidae) contain species which exploit phytophagous hosts, no
parasitic wasps are known to be predaceous as adults. Thus in
courtship, the delicate task of the female needing to discriminate
potential mate from potential food (such as in mantids or spiders)
does not arise, and the oftimes elaborate courtship behaviors observed cannot be attributed to this function.

Despite a large number of casual and fragmentary observations,
few detailed studies of mating and courtship in parasitic wasps
exist. As in other insects, however, parasitic wasp courtship consists of a series of highly specific reciprocal stimulus-response
sequences between the sexes. The subfamily Pteromalinae is the
best studied to date (van den Assem 1974), and probably the most
thorough available analysis of courtship in a parasitic wasp is
that of Barass (1960a,b, 1961) on *Nasonia vitripennis*. This small
pteromalid, a gregarious external parasite of fly puparia, exhibits
almost every component of courtship behavior which has been observed
to date in parasitic wasps. Insofar as known, this sequence in the
parasitic wasps generally includes most of the following: attraction,
recognition, orientation, wing vibration, antennation, head movements,
leg tapping, copulation, and post-copulatory grooming.

PATTERNS IN COURTSHIP: TWO EXAMPLES

To better appreciate the variety and complexity of parasitic
wasp courtship patterns, a close look at two actual cases seems
appropriate. Both of the following species have been the subject
of some opportunistic preliminary observations in our laboratory
during the past year, with 16 mm films through a dissecting micro-
scope allowing subsequent analysis of the courting sequence.

First, let us consider a cynipid, *Diastrophus nebulosus*. This
species is widespread in the eastern United States, and produces a
distinctive knot gall on wild blackberry, *Rubus* sp. Courtship in
D. nebulosus is relatively simple, and does not include a wing
vibration component. The male first makes antennal contact with

the female, who indicates receptivity by antennal position with
respect to her body plane. When non-receptive, a female holds her
antennae close to the substrate or under her body; a receptive female
holds her antennae roughly parallel to the substrate. Males often
mount non-receptive females and attempt to coax them into mating by
stroking their antennae up into a receptive position. If unsuccessful, they dismount and renew searching.

After mounting a receptive female, the male begins to antennate
vigorously, at the rate of 3.2 strokes per second, with a bobbing
motion of his head. He alternatively strokes the inner side of each
of the female's antennae with the outer side of his own corresponding antenna, the point of actual contact being at the sixth antennal
segment of the male. Remaining still, the female gradually elevates
her antennae as the male continues stroking. After antennating an
average of 60.5 seconds (range 50-68 sec., n = 6), the female exposes
her genital pocket, the male backs up, copulation follows, averaging
60.5 seconds (range 60-62 sec.). After copulation is completed, the
male dismounts and both sexes resume normal activity. Mated females
were not observed to mate again, but individual males often mated
repeatedly. *D. nebulosus* mating behavior is summarized in Fig. 1.

In Cynipidae, courtship has been studied in one other species,
Pseudeucoila bochei (van den Assem 1969). Although a parasitic
species, the basic courtship pattern is very similar to that of *D.
nebulosus* in its alternating antennal stroking (or "paddling"). But
the pace is slower, the number of strokes varying between 2.0 and 2.6
per second, and wing vibrating by males is an integral part of the
courtship. Copulation lasts 1 to 3 minutes in *P. bochei*, and females
are refractive to all subsequent mating attempts. In another species
of cynipid, *Synergus pacificus*, females are reported to mate several
times, with each copulation lasting only a fraction of a second
(Evans 1965).

For a second case study of parasitic wasp courtship, let us
consider a eulophid with a considerably more complex courtship
sequence. Recently, we have reared individuals of an apparently
undescribed *Aprostocetus* species (C. M. Yoshimoto, personal communication) from nests of a xylocopine bee, *Ceratina calcarata*, con-
structed in grass stems. *Aprostocetus* wasps develop as gregarious
internal parasites of the larvae and prepupae of these bees, the
females ovipositing through the stems to attack the host. Detailed
life history information is provided by Kislow (1975).

When 40 pairs of adults were observed after two days isolation
following emergence, an intricate courtship pattern was seen, as
summarized in Fig. 2. Upon successful mounting, the male begins
an integrated sequence of abdominal bending, metathoracic leg
extension and wing vibration. Beginning with all legs resting on
the female's thorax (hind legs on her fore wings), the male twists

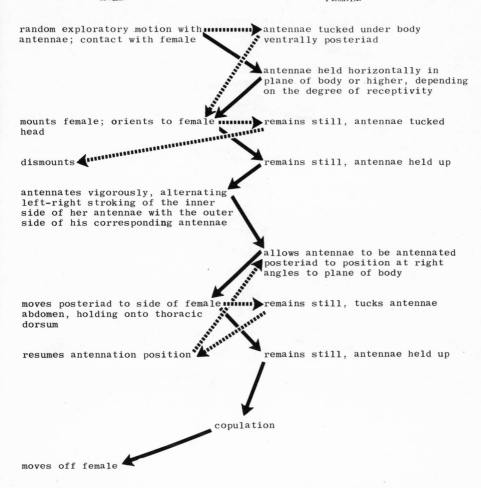

MALE   FEMALE

random exploratory motion with antennae; contact with female   antennae tucked under body ventrally posteriad

antennae held horizontally in plane of body or higher, depending on the degree of receptivity

mounts female; orients to female head   remains still, antennae tucked

dismounts   remains still, antennae held up

antennates vigorously, alternating left-right stroking of the inner side of her antennae with the outer side of his corresponding antennae

allows antennae to be antennated posteriad to position at right angles to plane of body

moves posteriad to side of female abdomen, holding onto thoracic dorsum   remains still, tucks antennae

resumes antennation position   remains still, antennae held up

copulation

moves off female

Fig. 1. Courtship sequence for *Diastrophus nebulosus* depicted as a reaction chain. Solid lines depict the normal or typical sequence: alternative directions are indicated by dashed lines.

his abdomen laterally, lowering its tip alongside the female's meso-thoracic leg until it is at an angle of 45° or more from the hori-zontal. Simultaneously, the male extends his metathoracic leg on the same side out and down, tapping it rapidly throughout the period of extension. Upon touching the substrate, it moves forward tapping on the female's middle tibia. Retracing these steps in reverse, the male returns to his original position above the female. However, one additional feature is present in the return; as the male's abdomen comes directly over the side of the female's abdomen, his

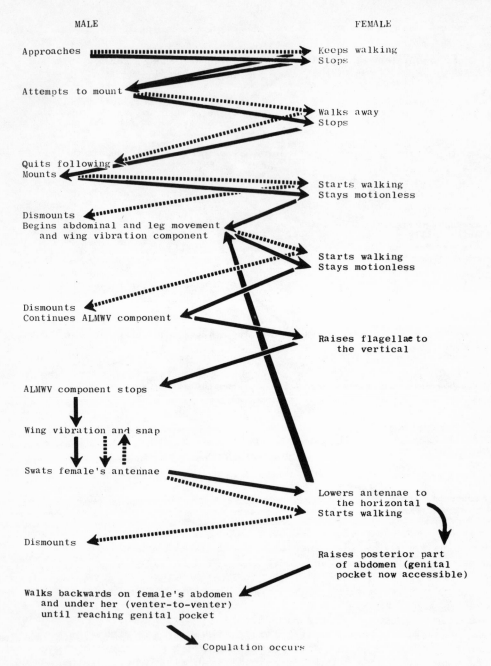

Fig. 2. Courtship sequence for *Aprostocetus* sp. depicted as a reaction chain. Solid lines depict the normal or typical sequence: alternative directions are indicated by dashed lines.

wings vibrate in a short burst. The vibration continues until his
abdomen is again back at its starting position. Then the male may
keep moving the abdomen toward the other side and perform the abdominal bending - leg extension - wing vibration cycle on that side.
The complete performance on one side takes an average of 2.2 seconds.
This cycle may proceed several times, usually with the male alternating sides. (Occasionally, we observed the cycle to proceed on
the same side of the female for 2-4 or more times in a row, but
this may have been due to the confining walls of the observation
chamber.)

For the most part, the receptive female remains motionless
during the male's courting. Her body is well-supported, and the
flagella of her antennae are held horizontally in front of her.
The male's flagella are held up at a 45° angle from the horizontal
and perpendicular to his longitudinal body axis. From time to time,
she raises her flagella until they are vertical. The male responds
first by halting the abdominal bending - leg extension - wing vibration cycle when his abdomen is mid-dorsal over the female. A burst
of wing vibration follows, ending with two simultaneous motions:
a) the snapping of the wings together, and b) the swinging of the
male's flagella forward and toward each other in the same plane as
they were held. The male's antennae contact ("swat") the female's
antennal flagella but do not push them into each other. The male
moves his flagella in an arc from the straight frontward position
at the time of swatting outward until the flagella form an angle
of 180° with each other, being roughly perpendicular to his longitudinal body axis when viewed from above. From there the flagella
retrace the arc anteriorly where the swat occurs. Following the
male swatting, the female may lower her flagella slightly, then
raise them to the vertical again, or may lower them to the horizontal
position again. The complete antennal cycle averages 1.1 second.
If the female does not lower her flagella, the male repeats the
"swatting" until she does lower them. The male then resumes the
sequence involving the abdomen, metathoracic legs and wings. This
sequence is repeatedly interrupted by the elevation of the female's
flagella; when the female's flagella lower, the sequence begins
again.

Aprostocetus courtship could be terminated at any time by
either sex. Males often walked off a female, sometimes returning
within 30 seconds to start another courtship sequence. Copulation
was observed in only three instances, lasting less than half a minute each time. The exact means by which the female signals her
readiness to copulate was not determined. No post-copulatory behavior was observed between the two sexes.

COURTSHIP AND REPRODUCTIVE ISOLATION

One of the functions of elaborate courtship behavior in animals is that of ensuring reproductive isolation between related species. In parasitic wasps, three studies of courtship behavior among closely related species have been made, all in the Chalcidoidea. Rao and DeBach (1969) and Khasimaddin and DeBach (1975) have analyzed court-ship behavior among several sibling species in the *Aphytis lingnanensis* complex (Aphelenidae), parasites of scales. Van den Assem and Povel (1973) have analyzed the similarities and differences among courtship patterns in three sympatric *Muscidifurax* species (Ptero-malidae), gregarious parasites of the housefly. Evans and Matthews (1975) have compared two *Melittobia* species (Eulophidae), ecological homologues attacking solitary wasps. In each case, courtship in the species studied turned out to differ distinctively, with behavioral species separations much more reliable than those based upon mor-phological criteria.

In *Aphytis*, two basic courtship patterns exist, distinguished by subtle differences in male attitude during courtship, position of female antennae, and male antennation patterns. The number of wing raises by the males and duration of courtship and post-copula-tory behaviors were found to be reliable characters for separating the sibling species.

The three species of *Muscidifurax* studied by van den Assem and Povel could be easily differentiated on the details of the antenna-tion movements and the relative duration of an antennal cycle. Average duration of both courtship and copulation also differed sig-nificantly among these species. Females of all three species invar-iably and perfectly discriminated between conspecifics and nonconspe-cifics.

Males of *Melittobia* have rather bizarre antennae which function as claspers for the female antennae, and the two species studied exhibit striking differences in the form of the antennation and leg kicking elements of courtship, as well as in relative courtship duration. *M. chalybii* males antennate females continuously, lifting the *meso*thoracic legs at regular intervals, and require about 10 minutes for the entire mating sequence. By contrast, *M.* sp. *A* males antennate in rapid alternation with *meta*thoracic leg pumping, and courtship duration is slightly less than two minutes.

THE ELEMENTS OF COURTSHIP

Pheromonal Attractants in Parasitic Wasps

Evidence for sex attractants produced by parasitic wasps is

widespread (van den Assem 1970, Bousch and Baerwald 1967, Obara and Kitano 1974, Jacobson 1972, Matthews 1974, Cole 1970, Rao and DeBach 1969, Vinson 1972). In virtually all cases, the female sex appears to be the source, with males orienting to increasing odor concentrations. Only for *Melittobia* has a male-produced sex attractant pheromone been implicated (Hermann et al. 1974), although a male pheromone functioning to quiet the female and make her more receptive is indicated in *Aphytis* (Rao and DeBach 1969), and it has been suggested that sweet odors emanating from various male braconids may be important in mating (Matthews 1974). Chemical characterizations of parasitic wasp pheromones have yet to be undertaken, hampered no doubt by the apparent absence of well-defined source glands in parasitic wasps (but see Buckingham 1968).

The few data which exist indicate that the pheromonal source may differ greatly from one wasp group to another. A series of experiments on the relative attractiveness of different female body parts of *Aphytis lingnanensis* (Rao and DeBach 1969) localized an active attractant factor in the thorax, apparently from an area near the base of the wings. For the braconid *Apanteles glomeratus*, the last abdominal segment of the female was found to be most active (Obara and Kitano 1974). All body regions of the female ichneumonid *Campoletis sonorensis* elicited courtship behavior by the male, and Vinson (1972) concluded that the pheromone was secreted through cuticular glands.

Male Wing Vibration During Courtship

In parasitic wasps, pulsed wing vibrations by males during courtship are the most conspicuous element of mating behavior, and seem to be of almost universal occurrence. Askew (1968) suggests that chalcid male wings may be more important for use in courtship than in dispersal. Facilitation of male orientation to a female's odor appears to be a primary function of such behavior, an interpretation given additional credence by the fact that in many species males will fan their wings upon being placed in containers where virgin females have previously been confined. Similar behavior is released when extracts of female bodies, or even the empty pupal skins of emerged females, are exposed to males (Obara and Kitano 1974, Cole 1970, Bousch and Baerwald 1967, van den Assem 1970, Vinson 1972). Using fine chalk dust, Vinson (1972) observed that wing fanning pulls odor-laden air over the male from front to rear, thereby permitting a directional orientation by the male to the odor source.

A suspicion arises, however, that wing fanning has additional roles in courtship and mating. Might the vibrations produced be important cues for the female? The question has been investigated in two unrelated species of chalcid wasps.

In normal courtship, male *Aphytis* repeatedly vibrate wings while mounted on the female prior to copulation. However, when Rao and DeBach (1969) carefully removed wings from both sexes of an *Aphytis* species, they still obtained 100% insemination in 10 pairings of these de-alated individuals. *Nasonia vitripennis* also vibrates its wings repeatedly during mounted precopulatory behavior, producing a sound shown to differ in frequency and pulse duration for normal and de-alated males (Miller and Tsao 1974). Some wingless males, in fact, failed to produce sound at all. When *Nasonia* females were paired with de-alated males, the offspring were predominantly of the male sex, which are produced from unfertilized eggs. However, some female progeny were also produced, indicating that at least a few successful matings may have occurred.

Neither of these experiments completely rules out the possibility that the sounds of courting males may have significance in species identification. The fact that such sounds are produced primarily during mounted precopulatory behavior suggests an additional interesting possibility. Such sounds may not necessarily be produced only by the wing fanning, but perhaps also by the associated thoracic muscle contractions, transmitted to females via the male's legs. If this hypothesis is valid, the placement of the male legs upon the female during courtship assumes new significance. In certain species (e.g., *Eupteromalus* sp., *Tetrastichus calamarius*, cited in van den Assem 1974), the males perform bouts of drumming of their front legs on the females' heads; these bouts may be important as acoustic signals as well.

Role of Antennae During Courtship

The role of the antennae in the mating sequence has been investigated by various authors (Grosch 1947, Obara and Kitano 1974, Vinson 1972, Rao and DeBach 1969). In general, antennectomized females are unresponsive to male courtship, and antennectomized males are unable to recognize and orient to females. Since courtship in many species includes sessions of males stroking female antennae with their own, it is likely that important tactile stimuli, or perhaps chemo-tactile stimuli involving surface pheromones, are involved. Vinson (1972) has suggested in addition that palpation of females by males serves to correctly orient the male to the female's body. This, of course, is followed by assumption of the proper mounted position for continued courtship.

In addition, the female's antennae are apparently important positional cues to courting males. Using dead female models fastened in different positions and exposed to courting male *Lariophagus*, van den Assem (1970) found that an upside down model with the antennae hidden caused males to mount and attempt courtship in a reversed position, the male orienting instead to the slightly

extended ovipositor (pseudo-antenna) of the model. Finally, females
often employ antennal positions to signal receptivity and nonreceptivity to males. Readiness to copulate following male courtship
stimuli is sometimes also indicated by stereotyped antennal position
changes (van den Assem 1974).

COURTSHIP STRATEGIES

A large percentage of female parasitic wasps are monandrous--
even when avidly courted by sexually stimulated males, an inseminated
female remains refractive to all subsequent copulatory attempts.
Moreover, it seems clear that control of mating receptivity rests
with the female. In chalcids, for example, receptivity to copulation
is apparently universally indicated by a characteristic postural
change in which the female raises her abdomen, thus exposing the
genital pockets; this behavior is often coupled with characteristic
antennal and head position changes (van den Assem 1974). Until the
genital pocket is exposed, it is impossible for the male to make
genital contact. Additionally, unreceptive females will often try
to move away and/or vigorously attempt to dislodge mounted males
with thrusts of their legs. Males of parasitic wasps, on the other
hand, are notorious for their non-discriminatory polygynous mating
behavior. Most will copulate repeatedly, and will mount anything
remotely resembling another female, including other males and host
pupal cells from which females have emerged. To a male, recently
mated females are just as attractive as are virgins. Moreover, not
infrequently several males will simultaneously attempt to court a
single female; while one is antennating, a second may be copulating
(van den Assem 1974).

Courtship strategies must also be affected by life history
constraints. For many parasitic wasps, the biological backdrop
against which courtship is acted out is that of an "island" breeding
structure, with random mating within almost-isolated subpopulations
(Hamilton 1967). Only one female usually parasitises a host, leav-
ing offspring of both sexes. Two additional facts are of significance. First, the sex ratio is usually female biased. Second,
neither sex tends to disperse until after mating, when females fly
off to search for new hosts. Thus, an individual can mate only
with a rather permanent set of neighbors. These are usually brothers and sisters, although because of occasional double parasitism
of a host, the resultant matings for the whole population in most
species are actually a mixture of many sibmatings and some out-
crosses. While the ability of a male to perpetuate his own genes
is inextricably tied to his success in inseminating females, the
reproductive success of the female is more directly related, instead,
to her host searching ability. The number of offspring a female
produces is limited by the amount of energy that can be mobilized
for egg production; she is not apt to have more offspring by repeated

matings than she does with one. In fact, because of the haplo-
diploid system of reproduction, an uninseminated female is still
able to produce offspring, although they will be exclusively of one
sex. However, should she fail to successfully find and parasitize
a host, her gametes would be wasted. In nature, then, the dispersal-
associated changes in behavior which occur immediately after copulation constitute an important part of the behavioral repertoire by
which monandry is maintained. In *Melittobia*, for example, we have
never found a virgin among those females which have left the host.

In such a situation, evolution would favor behaviors maximizing
the probability of female fertilization by the relatively few available males. Polygyny clearly must always be advantageous. However,
if male sperm is limited in supply, as observations by Simmonds
(1953) and Sekhar (1957) suggest, one might expect selection to
favor behaviors which decrease the likelihood that the polygamous
male spend his sperm needlessly upon repeated copulations with
females which have already been successfully inseminated. Restrict-
ing insemination to virgins increases the likelihood of maximally
successful reproduction for both sexes within the population, but
especially for the female whose monandrous behavior is thus clearly
beneficial to the species.

One would also expect selection to favor increased male persistence in courtship, for gains in male reproductive success could
be realized if initially unreceptive females might later be induced
to mate again. In certain Diptera, after bouts of oviposition
females become receptive again to male courtship, apparently as a
result of depletion of their stored sperm (Manning 1962). However,
as the sex ratio becomes increasingly skewed and the number of
females available relative to the male's maximal sperm capacity
increases, such an advantage probably tends to become relatively
less important. Receptivity and its relationship to male courtship
persistence in parasitic wasps has been thoroughly studied only in
the pteromalid genus *Muscidifurax* (van den Assem and Povel 1973).
Males of all three species persisted in courtship of unreceptive
(previously inseminated) females for significantly longer time than
they spent in courtship with virgin females, and for two species the
number of antennal cycles performed was significantly greater. However, the increased duration probably reflected only the slowed
responses of the nonreceptive females. Of significance is the fact
that although males mounted between 86 and 100% of the previously
inseminated females, they did not succeed in copulation in any case.

In connection with the maintenance of monandry, some little
noticed observations of van den Assem (1969, 1970) are of considerable interest. In both the cynipid *Pseudeucoila bochei* and the completely unrelated pteromalid *Lariophagus distinguendus*, not all male
courtship encounters with receptive virgin females led to copulation,
he found. After seemingly normal sequences of courtship behavior,

the male sometimes simply dismounted, even though the female had adopted the proper copulation posture with open genital pocket. Females stimulated to that point tended to remain in the copulation position for a time period roughly equivalent to the normal duration of a successful copulation. Thereafter, they resisted all further courtship attempts. When placed with hosts, these females produced exclusively male offspring.

In many parasitic wasps, it appears that the ability of the female to adopt a copulation posture, and hence to be fertilized, is only possible once, following the correct stimulation provided by a courting conspecific male. Identical stimuli, even in far greater quantity than needed for the primary effect, do not alter the subsequent nonreceptivity. This phenomenon has been termed "pseudovirginity" by van den Assem (1969), a confusing choice of words; "pseudo-insemination" would perhaps be a better term. Pseudo-insemination may be regarded as perhaps the ultimate error that a female might make. For a parasitic wasp, it is less serious, however, than it would be for vertebrates or even other insect groups, where failure to mate means no progeny at all. Uninseminated female wasps can still produce offspring--but they will be exclusively male, clearly a less than ideal situation.

Thus, the elaborate courtship behavior so commonly observed among parasitic wasps may often be a result of the fact that females characteristically will mate only once. Because of their monandrous behavior, there is strong selective pressure for a female to be "convinced", before signalling copulatory readiness, that her suitor is the proper species, fully viable, etc. A genetically foolish virgin female would be one who could readily be persuaded to adopt the copulation posture by any sexually stimulated male that happened along; the genetically wise would be particular and discriminating about potential mates. In addition, as the existence of pseudo-insemination indicates, once a female reaches a certain point in the courtship sequence, she is "triggered" into subsequence nonreceptivity whether or not copulation has actually occurred. Thus, an elaborate courtship also functions as a series of safeguards against the possibility of accidental triggering of a refractory state during encounters with individuals of the wrong species and/or sex.

In considering male polygamy and female receptivity, one additional possibility deserves more study than it has received to date. This is that females are somehow capable of discriminating between different males in relation to genetic differences or courtship vigor. Such phenomena have been investigated in only one species, *Nasonia vitripennis*. Studies by Grant et al. (1974) failed to find evidence for sexual selection (rare male advantage) by females of *N. vitripennis*. The question as to whether age and/or experience might alter the courtship vigor of *Nasonia* males was investigated by Barass (1960b). Aging was associated positively with a general slowing of four out of five courtship parameters; only the number of head movements per head series did not significantly change with increasing age.

While female choice must be regarded as the most significant selective force in the evolution of parasitic wasp mating systems, certain facets of male sexual strategy merit further consideration. Because a male's reproductive success can only be measured by the number of females he can successfully inseminate, competition among conspecific males is implicit. An "unbeatable" male sexual strategy would be to find and inseminate as many virgin females as possible before rival males.

One way to maximize the probability that an encountered con- specific female will still be virgin is to attempt to reach her as soon as she emerges or even before actual emergence. Thus, there would tend to be a selective premium upon development of behaviors enhancing copulation with females at increasingly earlier times in development. Ichneumonids of the genus *Megarhyssa*, the spectacular long-tailed wasps parasitic upon horntail larvae in wood, are of interest in this regard. Males congregate around sites of future female emergence and often attempt to insert their abdomens into bark crevices, sometimes to their full length. The possibility that such activities may result in fertilization of the female before emergence has been suggested by several authors (see review in Nuttall 1973), although generally discounted because of the greater length of the female abdomen as compared with that of the male. When three *M. nortoni* females which had been the cause of male aggregation were isolated upon emergence and placed on a host log where ovipos- ition occurred, a few individuals of both sexes subsequently emerged. Nuttall (1973) considered this proof that at least one female must have been successfully inseminated. An even more extreme situation exists in certain fig wasps (Agaonidae) where apterous males search out and copulate with pre-emergent females still in their pupal galls (Ramirez 1970). The telescoped abdomen of the male is inserted through a hole made in the female's gall; following female insemina- tion, males cooperatively chew a single emergence hole out of the fig, through which females subsequently escape.

While it is essential that male fig wasps cooperate in order for any of their genes to be perpetuated, this is not necessarily true in other groups. Mechanisms which serve to reduce competition from conspecific males also attempting to copulate assume consider- able importance in many species, particularly those which develop gregariously. Territorial behavior is one such mechanism. For example, in *Nasonia vitripennis*, a single host blowfly puparium produces an average of four males and 14 female wasps. Males emerge shortly before females and compete with each other for possession of the puparium. Having driven off rival males, the successful contender then mates with the virgin females as they emerge (King,

Askew and Sanger 1969).

Outright aggression between conspecific males is another such mechanism. For example, *Melittobia* is representative of what Hamilton has termed the ideal biofacies of extreme inbreeding and arrhenotoky. There is gregarious development of siblings, and females mate once with a brother immediately after eclosure. The flightless, blind males emerge first and are capable of mating many times, after which they die in the host cell where they were born. As in most other gregarious sib-mating parasitic wasps, there is a pronounced economy in male production. Sex ratios are biased strongly in favor of females, with males constituting less than 5% of offspring, but rarely are no males at all produced. Many colorful descriptions (summarized in Dahms 1973) indicate that vigorous fighting is commonplace among adult *Melittobia* males. Why should this be so?

In theory at least, a single male *Melittobia* could successfully inseminate all of his female clutch-mates. Accomplishing this would yield the maximum reproductive benefit a single male could expect to achieve, since the likelihood of unfertilized females from other clutches dispersing to find him is remote. By allowing brothers to share in the task of insemination, a male would, in effect, be sacrificing a considerable amount of his own reproductive potential.

In cases with extremely female-biased sex ratios, selfishness manifested in aggressive behavior is likely to reap increasingly greater rewards, particularly as the absolute number of males available for copulation declines. For example, if 100 females and five males are produced in a hypothetical clutch and the five adult males somehow eliminate one of their number prior to his copulating, then each of the surviving males obtain a 25% increase in females available to them for mating. If only two of the five males survive, their benefit increases by 150% each--a tremendous improvement in potential reproductive success to the survivors.

Interestingly, in the two *Melittobia* species we have studied, the form of male competition is manifested in strikingly different ways. In the course of obtaining comparative development times for *Melittobia* male pupal stages, Campbell (1973) segregated male pupae developing on a given host into separate capsules for observation. To his surprise, 1 to 3 days later only one intact male typically remained, surrounded by dismembered parts of other males. This situation was particularly pronounced by *M*. sp. *A* (Table 1). What appeared to be happening was that the first male to emerge typically eliminated his brothers by systematically decapitating them, either just prior to their eclosion or very soon afterwards. In undisturbed cultures allowed to die out after one generation, total adult population counts revealed that nearly every replicate contained one or more male heads, as well as various other male body parts. No decapitated or dismembered females were noted in any

Table 1. Comparison of observed and real sex ratios expressed as average percent of males in two species of *Melittobia* at three rearing temperatures. Real sex ratios were determined by adding the number of male heads to the number of intact males counted in expired one-generation cultures and recalculating. (Based on data in Campbell 1973).

Temp. C°	M. sp. A				M. chalybii			
	n	sex ratio observed	sex ratio real	% male progeny killed	n	sex ratio observed	sex ratio real	% male progeny killed
31	25	2.73	3.37	19.1	24	3.09	3.12	0.9
26	27	1.68	2.85	41.1	24	2.65	3.00	11.9
21	29	2.79	4.60	39.4	25	3.49	3.74	6.7

culture, suggesting that the dismembered males had been specifically
selected out for physical attack. We now believe that a majority
of decapitations occur in the very late pupal stages, just prior to
emergence, which means that no fight in the usual sense of the word
occurs; essentially the situation is one of the murder of helpless
rivals. Interestingly, when one observes interactions between conspecific *adult* males, one finds that this species shows far less
aggression than does *M. chalybii*. Not only is the fighting less
vigorous, but males of *M. sp. A* sometimes adopt an inert or submis-
sive posture when contacted by another male; pulling all appendages
tightly against the body, they assume a position which seems to
render a certain degree of immunity to further attack by the aggressor. An aggressor tends to lose interest in an "inert" male, and
soon abandons it without injury.

Melittobia chalybii adopts a different strategy. It is apparent
that adult male mortality in *M. chalybii* is less than in *M. sp. A*
at all rearing temperatures (Table 1), although some pupal decapitations may still occur. However, when conspecific adults encounter
one another, the interactions are intense. Upon contact, one or
both individuals will attempt to grasp some part of the other's body
with the mandibles; generally this is accompanied by much flailing
of legs and tumbling about. The outcome of such male contests,
although generally not culminating in death, usually involves some
form of disablement, such as loss of an appendage. The "inert"
postural response to aggression was never observed in this species.

In summary, while competition among males clearly occurs in
both species, and results in fewer males being available for mating,
the behaviors involved differ strikingly. *M. chalybii* appears to
be the classic pugnacious species of the literature (though confusion
regarding specific identities necessitates caution; for summary,
see Evans and Matthews 1975). On the other hand, competition among
M. sp. A males primarily takes the form of murder of pre-emergent
males by newly emerged adult males. This latter strategy would
appear to be the more effective method for reducing competition,
on a theoretical basis as well as an observational one. It has the
advantages of being more predictable and less hazardous, since when
both contestants are adults there is always the possibility that they
might incapacitate each other. Incidentally, inspection of Table 1
reveals that mortality through male aggression can have a considerable effect on the "observed" sex ratios. Hartl's (1971) suggestion
that sex ratios below predicted optima might be due to newly emerging males killing off other males seems supported by the data on
Melittobia.

The possibility of a nutritional benefit resulting from male
aggression has not been considered. Two lines of evidence, however,
suggest that killed individuals may provide an important source of
nutrition for the surviving victor. One is that pre-emergent males

are often dismembered. A more direct bit of evidence, however, is that on several occasions we have observed male *M. chalybii* attacking virgin females presented to them in the mating chambers, rather than attempting to mount. When this occurs, the male usually tears a hole in the female's abdomen and chews vigorously on her for several minutes. Other workers have consistently stated that male *Melittobia* do not feed, but our repeated observations of this behavior suggest that it may be a rather common phenomenon. If murder of one female provides enough nourishment to manufacture sufficient sperm to inseminate many more females, then it must be viewed as adaptive in that the male's, as well as the female's, reproductive success will be increased. However, if male feeding is important, then it would seem more selectively advantageous for males to feed on other males rather than upon females. It may be that male aggressive behavior has additional, or even different, evolutionary roots than usually postulated.

SUMMARY

Courtship may be defined as any behaviors between conspecific individuals of opposite sex which facilitate mating. Despite an incredible diversity of life histories, parasitic wasp adults are characteristically not carnivorous, and their oftimes elaborate courtships cannot have evolved to signal distinctions between mate and food, as is true for predominantly predaceous groups.

As two examples of courtship sequences, preliminary observations on a cynipid, *Diastrophus nebulosus*, and a eulophid, *Aprostocetus* sp., are presented. Courtship in *D. nebulosus* is rather simple, consisting of the male alternately stroking the female's antennae with his antennae at the rate of 3.2 strokes per second. No wing vibration component was present. *Aprostocetus* exhibit an intricate courtship pattern with two main phases. The basic phase consists of an integrated sequence of abdominal bending, metathoracic leg extension and tapping, and wing vibration; the whole phase lasts 2.2 seconds on each side. During this phase the antennae of the male do not contact those of the female. Interspersed within this sequence is a second pattern which is initiated when the female moves her antennae upwards. For this phase, the male changes to an antennal cycle requiring 1.1 second, in which his flagella move in an arc from straight ahead (where they "swat" the female flagella) to straight out sideways (at a 90° angle to body axis), then return to meet in front of the male's head. Immediately prior to the antennal contact, the male wings vibrate and then snap closed together. Diagrammatic summaries of the courtship reaction chains for these two species are provided in Figs. 1 and 2.

A brief review of three studies comparing courtship behavior among sets of closely related sympatric chalcid species is made;

distinctive differences in courtship pattern apparently serve to
ensure reproductive isolation between the species. Elements of
courtship as found in parasitic wasps are considered and selected
evidence pertaining to these behaviors is summarized together with
some functional interpretations. Sex pheromones may be produced by
either sex but none are yet chemically characterized for parasitic
wasps. The nearly universal occurrence of male wing vibration during
courtship facilitates proper orientation to females; evidence for
their additional role as an auditory signal during courtship in
some species is also considered. Antennation of females by males
during courtship probably provides important tactile or chemo-tactile
signals, since antennectomized wasps are generally unable to recognize and properly orient to the opposite sex. By means of characteristic positional changes, female antennae play an important role
in signalling readiness to copulate.

   A large percentage of parasitic wasp females are monandrous,
while males are characteristically polygynous. Sex ratios tend to
be female biased, often strongly so, and mating occurs prior to
dispersal. Monandry is viewed as a mechanism for maximizing the
distribution of available male sperm among the more numerous females
with minimal wastage. Thus, elaborate courtships so common in gregarious parasitic wasps may have evolved in response to the over-
whelming need for the female to make the correct choice the first
time. The occurrence of "pseudo-insemination", where the stimulus
of male courtship triggers the female into subsequent refractoriness
independent of actual insemination, furnishes a further pressure
toward elaborate courtship sequences which then function as a fail-
safe system.

   Since a male's reproductive success can only be measured by the
number of females he successfully inseminates, competition between
conspecific males for available females is implicit. Such competition may be manifested through ever earlier copulation in relation
to female emergence, territorial behaviors or outright aggression.
A notable example of male pugnacity occurs in the gregariously
developing genus *Melittobia*. *M. chalybii* males are pugnacious in
the classical sense, and fight one another vigorously. In contrast,
M. sp. *A* males are relatively non-aggressive toward one another as
adults; however, the first-emerging male evidently selectively
decapitates his brothers just prior to their emergence, thus effectively eliminating much subsequent competition for the right to
inseminate females. The possibility that aggression may have evolved
as a means for obtaining adequate nutrition to successfully court
and inseminate large numbers of females is also discussed.

ACKNOWLEDGEMENTS

It is a pleasure to acknowledge a number of students who have

assisted in gathering data reported herein and in the collection
and maintenance of the wasps involved. Thanks are extended
especially to Hilton C. Bruch, Thomas R. Campbell, Curtis E. Dunn,
Lisa D. Hermann and Carol L. White. I am particularly indebted to
Dr. David A. Evans, who extensively observed the *Melittobia* behavior
and served as a source of ideas and constructive criticism during
his sabbatical year at the University of Georgia. Dr. Carl M.
Yoshimoto of the Canadian National Collection in Ottawa kindly
identified the wasps, vouchers of which have been deposited in
collections both there and in the Entomology Department Museum
at the University of Georgia.

LITERATURE CITED

Alexander, R. D. 1964. The evolution of mating behaviour in
    arthropods. p. 80-92. *In* K. C. Highnam, ed. Insect
    Reproduction. Symp. R. Entomol. Soc. Lond.
Askew, R. R. 1968. Considerations on speciation in Chalcidoidea
    (Hymenoptera). Evolution 22:642-645.
Askew, R. R. 1971. Parasitic insects. American Elsevier, New
    York. 316 p.
Assem, J. van den. 1969. Reproductive behaviour of *Pseudeucoila
    bochei* (Hymenoptera: Cynipidae). I. A description of court-
    ship behaviour. Neth. J. Zool. 19:641-648.
Assem, J. van den. 1970. Courtship and mating in *Lariophagus
    distinguendus* (Forst.) Kurdj. (Hymenoptera, Pteromalidae).
    Neth. J. Zool. 20:329-352.
Assem, J. van den. 1974. Male courtship patterns and female
    receptivity signals of Pteromalinae (Hymenoptera, Pteromalidae),
    with a consideration of some evolutionary trends and a comment
    on the taxonomic position of *Pachycrepoideus vindemiae*.
    Neth. J. Zool. 24:253-278.
Assem, J. van den, and G. D. E. Povel. 1973. Courtship behaviour
    of some *Muscidifurax* species (Hym., Pteromalidae): a possible
    example of a recently evolved ethological isolating mechanism.
    Neth. J. Zool. 23:465-487.
Barass, R. 1960a. The courtship behaviour of *Mormoniella vitri-
    pennis* Walk. (Hymenoptera, Pteromalidae). Behaviour 15:
    185-209.
Barass, R. 1960b. The effect of age on the performance of an innate
    behaviour pattern in *Mormoniella vitripennis* Walk. (Hymenoptera,
    Pteromalidae). Behaviour 15:210-218.
Barass, R. 1961. A quantitative study of the behaviour of the male
    Mormoniella vitripennis (Walker) (Hymenoptera, Pteromalidae)
    towards two constant stimulus-situations. Behaviour 18:288-311.
Bousch, G. M., and R. A. Baerwald. 1967. Courtship behavior and
    evidence for a sex pheromone in the apple maggot parasite,
    Opius alloecus. Ann. Entomol. Soc. Amer. 60:865-866.

Buckingham, G. R. 1968. Pygidial glands in male *Opius*. Ann. Entomol. Soc. Amer. 61:233-234.

Campbell, T. R. 1973. Competition between two parasitoids, *Melittobia chalybii* Ashmead and *Melittobia* species A (Hymenoptera: Eulophidae). M. S. Thesis, University of Georgia. 54 p.

Clausen, C. P. 1940. Entomophagous insects. McGraw-Hill, New York. 688 p.

Cole, L. R. 1970. Observations on the finding of mates by male *Phaeogenes invisor* and *Apanteles medicaginis*. Anim. Behav. 18:184-189.

Dahms, E. 1973. The courtship behaviour of *Melittobia australica* Girault, 1912 (Hymenoptera: Eulophidae). Mem. Queensland Mus. 16:411-414.

Doutt, R. L. 1959. The biology of parasitic Hymenoptera. Annu. Rev. Entomol. 4:161-182.

Doutt, R. L. 1964. Biological characteristics of entomophagous adults. p. 145-167. *In* P. DeBach, ed. Biological control of insect pests and weeds. Chapman and Hall, London.

Evans, D. 1965. The life history and immature stages of *Synergus pacificus* (McCracken and Egbert) (Hymenoptera: Cynipidae). Can. Entomol. 97:185-188.

Evans, D. A., and R. W. Matthews. 1975. Comparative courtship behaviour in two species of the parasitic chalcid wasp *Melittobia* (Hymenoptera: Eulophidae). Anim. Behav. In press.

Grant, B., G. A. Snyder, and S. F. Glessner. 1974. Frequency-dependent mate selection in *Mormoniella vitripennis*. Evolution 28:259-264.

Grosch, D. S. 1947. The importance of antennae in the mating reaction of male *Habrobracon*. J. Comp. Physiol. Psychol. 40:23.

Hamilton, W. D. 1967. Extraordinary sex ratios. Science 156: 477-488.

Hartl, D. L. 1971. Some aspects of natural selection in arrhenotokous populations. Amer. Zool. 11:309-325.

Hermann, L. D., H. R. Hermann, and R. W. Matthews. 1974. A possible calling pheromone in *Melittobia chalybii* (Hymenoptera: Eulophidae). J. Georgia Entomol. Soc. 9:17.

Jacobson, M. 1972. Insect sex pheromones. Academic Press, New York. 382 p.

Khasimuddin, S., and P. DeBach. 1975. Mating behaviour and evidence of a male pheromone in species of the genus *Aphytis* Howard (Hymenoptera: Aphelinidae). Manuscript.

King, P. E., R. R. Askew, and C. Sanger. 1969. The detection of parasitized hosts by males of *Nasonia vitripennis* (Walker) (Hymenoptera: Pteromalidae) and some possible implications. Proc. R. Entomol. Soc. Lond. Ser. A. 44:85-90.

Kislow, C. J. 1975. The comparative biology of two species of little carpenter bees, *Ceratina strenua* and *C. calcarata* (Hymenoptera, Xylocopinae). Ph.D. Dissertation, Univ. of Georgia.

Manning, A. 1962. A sperm factor affecting the receptivity of
    Drosophila melanogaster females. Nature 194:252-253.
Matthews, R. W. 1974. Biology of Braconidae. Annu. Rev. Entomol.
    19:15-32.
Miller, M. C., and C. H. Tsao. 1974. Significance of wing vibration
    in male *Nasonia vitripennis* (Hymenoptera, Pteromalidae) during
    courtship. Ann. Entomol. Soc. Amer. 67:772-774.
Nuttall, M. J. 1973. Pre-emergence fertilization of *Megarhyssa
    nortoni* (Hymenoptera: Ichneumonidae). New Zealand Entomol.
    5:112-117.
Obara, M., and H. Kitano. 1974. Studies on the courtship behavior
    of *Apanteles glomeratus* L. I. Experimental studies on releaser
    of wing-vibrating behavior in the male. Kontyu 42:208-214.
Orians, G. H. 1969. On the evolution of mating systems in birds
    and mammals. Amer. Nat. 103:589-603.
Ramirez, B. W. 1970. Taxonomic and biological studies of Neotrop-
    ical fig wasps (Hymenoptera: Agaonidae). Univ. Kans. Sci.
    Bull. 49:1-44.
Rao, S. V., and P. DeBach. 1969. Experimental studies on hybridi-
    zation and sexual isolation between some *Aphytis* species
    (Hymenoptera: Aphelinidae). I. Experimental hybridization
    and an interpretation of evolutionary relationships among the
    species. Hilgardia 39:515-554.
Sekhar, P. S. 1957. Mating, oviposition, and discrimination of
    host by *Aphidius testaceipes* (Cresson) and *Praon aguti* Smith.
    primary parasites of aphids. Ann. Entomol. Soc. Amer. 50:
    370-375.
Simmonds, F. J. 1953. Observations on the biology and mass-breed-
    ing of *Spalangia drosophilae* Ashm. (Hymenoptera, Spalangiidae),
    a parasite of the frit-fly, *Oscinella frit* (L.). Bull. Entomol.
    Res. 44:773-778.
Spieth, H. T. 1974. Courtship behavior in *Drosophila*. Annu.
    Rev. Entomol. 19:385-405.
Townes, H. 1969. The genera of Ichneumonidae, part I. Mem. Amer.
    Entomol. Inst. 11. 300 p.
Trivers, R. L. 1972. Parental investment and sexual selection.
    p. 136-179. *In* B. Campbell, ed. Sexual selection and the
    descent of man, 1871-1971. Aldine, Chicago.
Vinson, S. B. 1972. Courtship behavior and evidence for a sex
    pheromone in the parasitoid *Campoletis sonorensis* (Hymenoptera:
    Ichneumonidae). Environ. Entomol. 1:409-414.
Weld, L. H. 1952. Cynipoidea (Hym.) 1905-1950, being a supplement
    to the Dalla Torre and Kieffer Monograph--The Cynipidae *in* Das
    Tierreich, Lieferung 24, 1910 and bringing the systematic
    literature of the world up to date, including keys to families
    and subfamilies and lists of new generic, specific and variety
    names. 351 p. Privately printed.
See author's note p. 224.

REPRODUCTIVE STRATEGIES OF PARASITOIDS

Peter W. Price

Department of Entomology, University of Illinois

Urbana, Illinois 61801, U.S.A.

Réaumur (1738) was perhaps the first biologist to express
wonder at the great diversity of fecundities seen among insects.
Although this diversity is better documented today than in Réaumur's
time, we are only slightly closer to understanding why it exists,
or why related species differ in the number of eggs they produce.
For parasitoids, knowledge of the selective factors acting on egg
production and the results of natural selection are necessary for
a full understanding of the coevolution of host and parasitoid,
host and parasitoid population dynamics, parasitoid community
ecology, and for the development of a predictive science in biological control using introduced parasitoids.

So far I have examined the selective factors operating on
parasitoids in the family Ichneumonidae (Hymenoptera). First the
mean values for ovariole number per ovary for species in each of
13 subfamilies were used with the assumption that ovariole number
was closely correlated with fecundity (Price 1973a). Although
this analysis demonstrated the existence of patterns in fecundity,
it was too general to answer questions such as "why do species
attacking the same stage of different hosts have different ovariole
numbers?" A more detailed study on ten species of ichneumonid
which attacked one host species accounted for most of the differences
in ovariole numbers between them (Price 1974). However this inten-
sive study lacked the generality that was appealing in the former
effort. Here I attempt to fill some of the middle ground by using
data on a moderate number of ichneumonid species, with moderately
well known life histories, and which collectively attack a great
variety of hosts in varied microhabitats. Using similar arguments
I also extend the explanation of fecundity differences to the large
family of parasitoid flies, the Tachinidae (Diptera).

DETERMINANTS OF PARASITOID FECUNDITY

Many factors are likely to influence the number of eggs a female is able to lay in a lifetime, and therefore the evolution of ovary structure for egg manufacture. These factors may be divided into those that affect the probability of discovering the host and ovipositing in it and those that determine the probability of survival once the egg is laid. These factors are closely linked. For example, a very low probability of discovering a host would lead to a low fecundity, but since the host is hard to find survival is likely to be high so the few eggs that are laid produce a relatively high number of reproductives in the next generation. Conversely where probability of discovery is high, probability of survival is likely to be low and high fecundity in parasitoids will be selected for.

In the first category the abundance of the host is important. The average absolute abundance from year to year of the host will certainly influence the number of eggs a parasitoid can lay. Also the relative abundance of the developmental stages of the host within one generation decreases because of mortality so parasitoids which attack early stages should find more hosts than those that attack later stages. The dispersion pattern of hosts must be influential as clumped or colonial hosts may be more easily discovered, although coloniality may provide a degree of protection. In fact any sort of defense mechanism in the host will essentially reduce the possibility for laying eggs. Physically protected hosts in burrows, mines or webbing reduce accessibility for parasitoids, and slow down the oviposition process, so fewer total eggs may be laid by a female attacking these types. Finally female parasitoids may not oviposit on the host but on its food, a behavior that is common in the Tachinidae, so the probability of discovering the host is greatly reduced, although probability of survival is also reduced. Here females should be extremely fecund since, contrary to the general cases listed above, both discovery and survival are precarious for any one individual.

The probability of survival once in the host can be estimated by the host survivorship curve since the parasitoids will be subject to similar mortalities to their hosts. Since these curves decline with host age the later the stage attacked the greater will be the probability of survival and the lesser the need for high fecundity. Highly protected hosts presumably survive better than exposed hosts so parasitoids on the former host type are likely to be less fecund. Internal eggs or larvae will probably survive better than those that are external. Thus we should expect an increase in fecundity adopted in conjunction with external stages, as in the Tachinidae, or a limitation of this type of oviposition and feeding to less vulnerable stages of the host, as we shall see in the Ichneumonidae.

Nutrition of parasitoid larvae and adults influences fecundity (e.g. Bracken 1966, Leius 1961) but this factor is more likely to account for differences within a species rather than differences between species.

HOST SURVIVORSHIP CURVES

Of all the factors influencing the evolution of average fecundity levels the easiest to examine is the effect of host survivorship on the probabilities of discovery and survival for parasitoids attacking different stages. In addition, use of survivorship curves has enabled an explanation of much of the differences in potential fecundity among ichneumonid parasitoids (Price 1973a, 1974). An examination of the general features of survivorship curves derived from natural populations of insect herbivores is therefore worthwhile. Data used in the construction of Fig. 1 were derived from the herbivores and sources provided in Appendix 1.

The available data suggest that there are two major types of survivorship curve among herbivorous insects (Fig. 1). Type A curves show that 70% or more of total mortality occurs by the mid larval or nymphal stage. Five of the 11 species in this type live in exposed and presumably vulnerable situations. For those that live in more protected sites establishment of the first instar is likely to be difficult. For example Waloff (1968) found in *Sitona regensteinensis* that most mortality occurred because the first instar larvae failed to reach root nodules in which they feed. Thus in Type A survivorship curves mortality rates are high in early stages of development and low in later stages. The only data on hemimetabolous insects belong here. In contrast Type B curves indicate high mortality rates in later larval stages, but 40% or less of mortality occurs by the mid larval stage. The seven species representing this type are holometabolous and most are protected, either by the site of feeding (five species) or by effective colonial defense (one species). Presumably adults suffer higher mortality than in the A type during placement of eggs in secure positions. Only *Pieris rapae* larvae feed in an exposed position, and here the cultured environment may provide a refuge from many natural mortality factors. Although most species fit into the Type A or B category, the survivorship curves for *Coleophora serratella* and *Leucoptera spartifoliella* indicate that a continuum of shapes probably exists between the types identified in Fig. 1.

From these generalized herbivore survivorship curves some predictions may be made on the relative probability of discovering hosts at different stages of development, and the relative probability of surviving in hosts once eggs or larvae have been deposited. Probability of discovery will decline in the same way as abundance of hosts declines (Fig. 2), although as dispersion tends to change

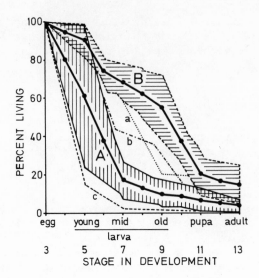

Fig. 1. General trends in survivorship curves for 19 species of
herbivorous insects listed in Appendix 1. Shaded areas indicate
the zones which include all survivorship curves in the group indicated, the species of which are listed in Appendix 1. Dots are mid
points between limits of the shaded areas, on the vertical scale,
and the heavy lines joining these indicate the general characteristics of Type A and B survivorship curves. Survivorship curves
end at the early adult stage. Presumably Type B species suffer
greater adult mortality than Type A species. Part of the survivorship curves for three species not included in the shaded areas are
given separately: a, *Coleophora serratella*; b, *Leucoptera spartifoliella*; c, *Lepidosaphes ulmi*. The numerical designation of host
developmental stage was used for regression and for easy reference
to subsequent figures.

in some cases from clumped to random the probability will decrease
more rapidly than indicated. The relative probability of surviving
in a host may be calculated from the host survivorship curve by
assuming a probability of 1.0 if the parasitoid leaves its progeny
in well developed pupae (Stage 12, Figs. 1 and 2). Thus, since
parasitoids in the Ichneumonidae and Tachinidae typically emerge
from the later stages of development the probability of survival
in the host decreases rapidly as earlier stages of the host are
attacked (Fig. 2).

These changes in relative probability, and the differences seen
in Type A and Type B host species will influence patterns in fecundity of parasitoids. Probabilities of survival in Type A and B
curves converge for species attacking late stages of the host

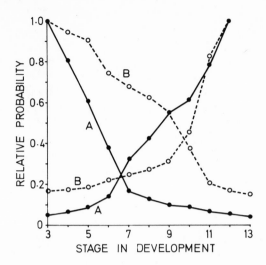

Fig. 2. The probabilities of discovering hosts and surviving in
them. The probabilities of discovering a host for Type A and B
hosts relative to the probability of discovering a host in the egg
stage (3) which is designated as p = 1.0. The probabilities of
surviving in a host relative to the probability at the late pupal
stage (12) which is designated as p = 1.0. This assumes that
parasitoids emerge from stage 12.

(11-12) so we should expect only small differences in fecundity
between these species. However parasitoids attacking earlier
stages (3-10) will experience very different probabilities of
survival according to the type of host they attack. Therefore we
should expect considerable differences in fecundity among species
attacking these stages. These differences should be greater than
predicted by the probability curves since these are conservatively
estimated from the mid points in the range of survivorship curves
and not from the extreme values. The double jeopardy of those
species that oviposit on host food, which usually attack early
stages of the host, may also be predicted. Probability of host
discovery is low and probability of survival is also low, so the
product of these probabilities will determine the evolution of
levels of fecundity. If the probability of discovering a host
is 0.1 for these species and we see from Fig. 2 that the probability of survival is also about 0.1, fecundity must be 100 times
larger than that in a species attacking the late pupal stage.
We will see in the next section of this paper that fecundities
in ichneumonid and tachinid parasitoids range within these two
orders of magnitude. Those species that attack early host stages
may be regarded as r strategists and those that attack late stages
as K strategists (cf. Price 1973b, D. C. Force and R. R. Askew

in this volume).

OVARIOLE NUMBER AND FECUNDITY

In the past ovariole number per ovary has been used as an estimator of relative fecundity in Ichneumonidae (Price 1973a, 1974), but the validity of this relationship was then only assumed. A survey of published records that include both fecundity and ovariole number for members of the families Ichneumonidae and Tachinidae now indicates that there is indeed a correlation, and that it is linear (Fig. 3 and Appendix 2). The relationship appears to be the same for members of both families.

From the probabilities of survival given in Fig. 2 it is possible to predict the number of eggs a female would require if she attacked an earlier stage of the host in order to achieve the same number of surviving progeny as another female that attacks stage 11 or 12. Only the fecundity of this latter female cannot be predicted. In the Ichneumonidae typical females attacking stage 11 average about 40 eggs in a lifetime (see species marked with dagger in Appendix 2). Using this figure all other expected fecundities have been calculated, and from these the number of ovarioles per ovary were read from Fig. 2. Thus fecundity and number of ovarioles per ovary are predicted for ichneumonid females attacking all stages of the host (Table 1). Note that females attacking late host stages of both Types A and B species have similar fecundities and ovarioles per ovary, and relatively large changes in probability of survival of progeny have a small influence on egg and ovariole numbers (cf. Fig. 2 and Table 1). The actual numbers observed should show a small variance. By contrast, females that attack early host stages of Types A and B hosts have very different fecundities and relatively small changes in the probability of survival of progeny influence greatly the number of ovaries required to equalize survival. Observed egg and ovariole numbers should show large variance among species attacking early stages. We should also predict that parasitoids attacking early host stages would require fewer eggs if attacking a B type host than a similar species which attacks an A type host (e.g. 195 eggs vs. 572 eggs). Since fecundity and numbers of ovarioles per ovary are related in a linear fashion only ovariole number will be discussed in the remainder of this paper.

ADAPTATIONS OF REPRODUCTIVE ORGANS AND LARVAE

The adaptations in the structure of reproductive organs associated with high and low fecundity have been examined for the Ichneumonidae (Price 1973a and references therein). In general a highly fecund female has many ovarioles per ovary, as shown in

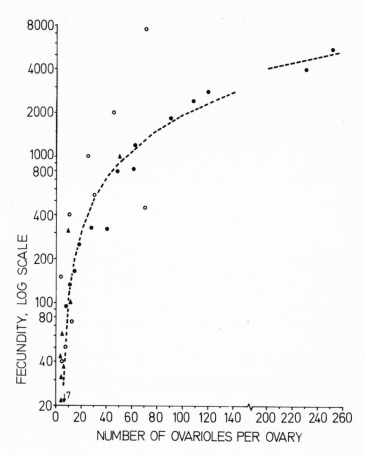

Fig. 3.   The linear relationship between number of ovarioles per
ovary and fecundity in Ichneumonidae (triangles) and Tachinidae
(circles) (see Appendix 2 for species, data and references).
Open circles are estimates made by Townsend, closed circles those
made by other authors. For all data points the regression equation
is Y = 97.2 + 21.53X, the correlation coefficient (r^2) is 0.57 and
p < 0.001. Since Townsend's estimates for the more fecund species
deviate widely from the general trend, whereas the estimates of
no other author do so, they may be inaccurate. With these data
points omitted the regression equation is Y = 20.66X - 96.69,
r^2 = 0.97 and p < 0.001. The line described by this equation is
shown in the figure.

Fig. 3, long lateral oviducts with high egg storage capacity, small
eggs, and a short ovipositor for piercing easily accessible hosts
(e.g. Fig. 4A). In contrast, those females with low fecundity have
few ovarioles per ovary, very short lateral oviducts too small to

Table 1. Predicted values for fecundity and ovarioles per ovary for
female ichneumonids attacking all stages of hosts of type A and B.

Stage of host attacked	No. of eggs required to equalize survival		No. of ovarioles per ovary	
	Type A host	Type B host	Type A host	Type B host
3	571.7	195.1	33.0	15.0
4	462.4	184.2	28.0	14.5
5	349.3	176.4	22.0	13.8
6	216.8	145.4	15.0	12.0
7	97.0	132.6	10.0	11.0
8	74.3	121.9	9.0	10.5
9	57.2	107.3	8.0	10.0
10	51.5	73.2	7.5	8.5
11	40.0	40.0	6.6	6.6
12	31.4	33.2	6.0	6.0

store a single egg, large eggs, a long ovipositor for reaching well
concealed hosts and a large uterus gland for lubrication of eggs
as they pass down the ovipositor (see Pampel 1914 for details)
(e.g. Fig. 4C). A poison gland and reservoir and alkaline gland
are common to all female ichneumonids as far as is known.

The Tachinidae were not preadapted for exploiting concealed
hosts as the appendages of the ovipositor have never been well
developed and a poison gland used in paralyzing hosts has not
evolved. However concealed hosts are still utilized because
eggs may be laid close to the entrance of a host burrow, or on
the soil above soil-dwelling hosts, and active planidia emerge
and burrow down to the host larva. Apart from this difference
the reproductive systems are similar. Among fecund females of
the Tachinidae the median oviduct is highly developed for egg
storage whereas the lateral oviducts remain small (e.g. Fig. 4B),
ovariole number is high and egg size is small. In females with
low fecundity the median oviduct is short and may contain only
one egg, the number of ovarioles is small and large eggs are
produced (e.g. Fig. 4D). The eggs of the majority of species
of Tachinidae may be incubated in the median oviduct which is
highly vascularized for the purpose and therefore acts as a uterus.
Incubation is rarely observed in the Ichneumonidae except in the
Tryphoninae (see Clausen 1940). Where eggs are laid in exposed
sites early hatching will reduce mortality, which is particularly
important in conjunction with the strategies evolved in the
Tachinidae as described later.

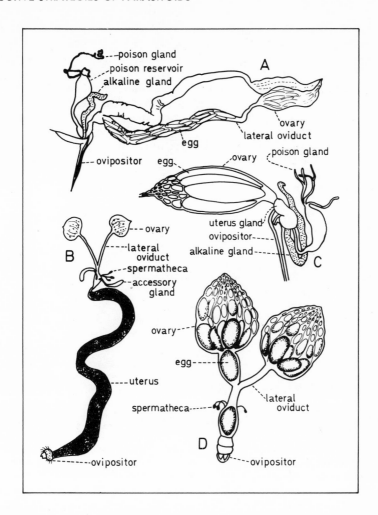

Fig. 4. Reproductive organs of female Ichneumonidae and Tachinidae.
A. *Enicospilus americanus*, Ichneumonidae (personal observation),
B. *Leschenaultia exul*, Tachinidae (after Bess 1936), C. *Trachysphyrus albatorius*, Ichneumonidae (after Pampel 1914), D. *Hyperecteina cinerea* Tachinidae (after Clausen et al. 1927). Reproductive systems are not drawn to the same scale. See text for further details.

Parasitoid larvae that hatch in or on the host have a color-less unsclerotized integument, no legs and reduced sensory apparatus. However some eggs are laid on the host's food, a habit adopted by very few Ichneumonidae (e.g. Eucerotini) a large number of Tachinidae and some chalcidoids, (e.g. Perilampidae). Eggs of some tachinids are ingested by the host, but in many species and in the other taxa first instar larvae hatch and a close evolutionary convergence to

the planidial habit is evident (see Tripp 1961, 1962, Clausen 1940).
Planidia are sclerotized, active and will attach to a passing host
or crawl until a host is discovered. The relatively low probability
of reaching a host when the egg is laid off the host body has resulted in strong selective pressure for an active first instar larva.
In fact the adaptive radiation of the Tachinidae has depended to
some extent on the use of this strategy, and has enabled the exploitation of concealed hosts that one would have thought were
available only to hymenopteran parasitoids. About 400 species of
tachinids have a planidial first instar out of the roughly 1300
species of Tachinidae in North America north of Mexico (D. M. Wood,
personal communication).

OVARIOLE NUMBER AND CHARACTERISTICS OF HOST ATTACKED

Ichneumonidae

Within 59 species of Ichneumonidae in which the ovariole number
and attack pattern could be ascertained with reasonable assurance
a trend was evident (Fig. 5, Appendix 3). As later stages of the
host were attacked, so the number of ovarioles per ovary declined
as found in previous studies. However, particularly for those
species attacking the egg stage the range was extreme. An exploration into their biology provides some reasons for this. Six of
the species (in group d Fig. 5) parasitize egg clusters, such as
those produced in a cocoon by spiders (Kerrich 1942, Townes and
Townes 1951, Townes 1969b) or in leaf rolls by weevils (Kerrich
1969). The parasitoid life cycle is completed on the egg mass.
Thus, the egg mass is more in the nature of a cocoon than an egg
both in physical properties and in dispersion. Cocoons protect the
host, they are usually concealed, less abundant than other stages
in the life cycle, and are more randomly dispersed than other
stages. Therefore we should expect the parasitoids attacking this
sort of host to have ovariole numbers similar to those which
attack cocoons containing pupae (stage 11), as indeed they do.

Another species which shows a large deviation from the general
trend is *Collyria calcitrator* (c Fig. 5). This species is unusually
host specific and oviposits in the well protected eggs of the wheat
stem sawfly and matures on the larvae (Salt 1931). The female parasitoid has a long ovipositor and in fact was classified in the boring
type (Bohrertypus) of apparatus by Pampel (1914), the only one in
this type that does not attack late stages of the host. Given that
the wheat stem sawfly survivorship curve is probably similar to that
for the European corn borer, *Ostrinia nubilalis*, which is known to
be a Type B host, the number of ovarioles per ovary (ten) does not
deviate excessively from the predicted 15 for females attacking
the eggs of a Type B host (see Table 1).

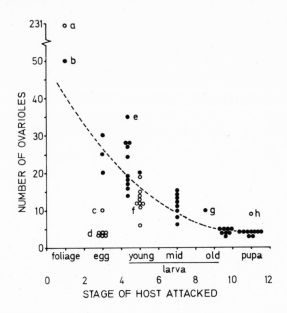

Fig. 5. The relationship between the number of ovarioles per ovary
in species of Ichneumonidae and the host stage attacked by each
species (see Appendix 3 for list of species and references). Points
at a and d were omitted from the regression for reasons given in
the text. The regression for the remainder of the points is Y =
$48.99 - 8.61X + 0.42X^2$, $r^2 = -0.73$, $p < 0.001$.

Members of the genus *Euceros* which lay their eggs on foliage
(a and b Fig. 5) have very high numbers of ovarioles per ovary, and
fecundities lie within the range predicted assuming a probability
of 0.1 that a planidium reaches a host. The more fecund species
(a in Fig. 5) was omitted from the correlation simply to avoid a
spurious negative value for number of ovarioles at stage 8 due to
the parabolic regression model used.

Four more taxa deviate considerably from the general trend.
Olesicampe lophyri (e in Fig. 5) is a parasitoid on the colonial
sawfly, Type B host, *Neodiprion swainei*. A lower fecundity for
such a species was predicted. This high number of ovarioles per
ovary may be possible because of the clumped dispersion pattern
of the host, enabling rapid oviposition, and the typically high
densities of this sawfly relative to other injurious forest insects
(cf. McLeod 1972). Likewise, *Lamachus lophyri* (g in Fig. 5),
another parasitoid on this sawfly has a higher ovariole number than
predicted. The group of species labelled f in Fig. 5 are all members
of the Banchinae, tribe Banchini, which emerge from larvae (stages
8 and 9), rather than from pupae. Thus the probability of survival

is underestimated in Fig. 2, and the ovariole number per ovary is consequently lower. Finally, *Agrothereutes tunetanus* (h in Fig. 5) has an ovariole number considerably higher than the other species. Several eggs per host are laid and the larvae develop gregariously, an uncommon habit in the Ichneumonidae, and one not shared by any other species attacking the pupal stage given in Fig. 5. Since several eggs must mature synchronously a larger number of ovarioles is required.

The predictions on ovariole number given in Table 1 can be compared to ovariole numbers read from the regression line given in Fig. 5. A slope (b) of one and a high correlation between the predicted and actual ovariole numbers would lend support to the claim that the host survivorship curve is the most influential determinant of parasitoid fecundity. If b = 1 there exists a 1:1 relationship between predicted and observed ovariole numbers. When the predicted values for a Type A host are compared with the mean values from Fig. 5 the correlation is high (r^2 = 0.99) and the slope is close to 1 (b = 0.80) (Fig. 6). When predictions from a Type B curve are used the correlation is lower (r^2 = 0.77) and the slope is too high (b = 2.22). When the mean values of predictions from the A and B Type hosts are compared with those read from the regression line the correlation is high (r^2 = 0.97) and the slope is nearer to one (b = 1.23). Since the type of host survivorship curve is unknown for the majority of hosts for parasitoids used in Fig. 5, this latter prediction is the most valid.

Tachinidae

The idea that egg production in Tachinidae has been evolutionarily adjusted to the probability of finding a host was expressed by Clausen (1940) and has received support since (Askew 1971). As far as I am aware the generality of this suggestion has not been tested rigorously. The probabilities of finding hosts can be ranked intuitively when the oviposition habits and the presence of a planidial stage are established (Fig. 7). Lowest probabilities are exhibited by species that lay eggs on food that do not hatch until ingested. The planidium increases the chance of finding a host and here those species that lay eggs at a host's tunnel entrance have a good chance of reaching a host as the planidium is directed in its search. Then larval hosts should be more easily discovered than adult hosts if the parent parasitoid oviposits on the host itself. However it seems that survival should be greater among parasitoids deposited within the host than those deposited on the host integument.

Townsend (1935, 1936a,b) provides the characteristics of 46 tribes of Tachinidae including ovariole number and attack pattern which could be used to test the relationship between probability

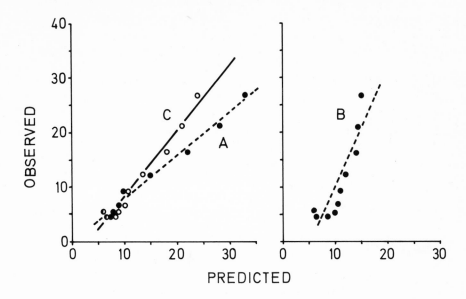

Fig. 6. The relationship between the "observed" number of ovarioles per ovary read from regression line in Fig. 5, and the predicted number given in Table 1, A, for parasitoids attacking a Type A host, B, for those attacking a Type B host and C, a mean of predictions from A and B type hosts. Equations are for A, $Y = 0.80X - 0.32$, $r^2 = 0.99$, $p < 0.001$; B, $Y = 2.22X - 12.61$, $r^2 = 0.77$, $p < 0.001$; C, $Y = 1.23X - 4.25$, $r^2 = 0.97$, $p < 0.001$.

of discovering a host and fecundity. However the variation within each tribe is great and the expected trend is very weakly developed. Thus details on individual species must be used to test the prediction adequately. In the literature 40 species were found with adequate information (Fig. 7, Appendix 4), and the correlation accounted for 66% of the variation. This correlation included three points for the Conopidae which oviposit on adult Hymenoptera. It seems that the parasitoid Sarcophagidae would also fit this trend although adequate information on mode of oviposition is hard to find. I conclude that the probability of discovering a host is the major determinant of fecundity in the Tachinidae and Conopidae with probability of survival as an additional contributing factor.

In order to promote further understanding of parasitoid reproductive strategies, in future studies it would be valuable to discover details of host survivorship and abundance, time of parasitoid attack relative to host development, probability of establishment in the host, and parasitoid fecundity and ovariole number.

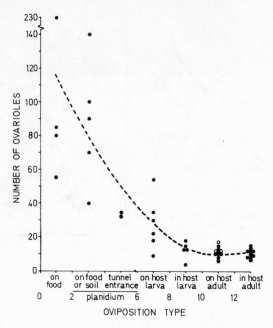

Fig. 7. The relationship between number of ovarioles per ovary in
species of Tachinidae (closed circles) and Conopidae (open circles),
and the type of oviposition practiced by the females (see Appendix 4
for list of species and references). The regression line is
described by the equation $Y = 138.29 - 22.77X + 1.00X^2$, $r^2 = -0.66$,
$p < 0.001$.

SUMMARY

     It is postulated that the major determinants of fecundity in
parasitoids are the probability of finding hosts and the probability
of survival once established on or in the host. Host survivorship
curves were summarized and, from these, predictions were made on the
fecundity and number of ovarioles per ovary required by parasitoids
for each stage of the host attacked. Fecundity is closely correlated
with ovariole number ($r^2 = 0.97$) in both Ichneumonidae (Hymenoptera)
and Tachinidae (Diptera). General features of ovary structure in
these families were described. An examination of 59 species of
Ichneumonidae supports the hypothesis. Predicted and observed ovariole numbers are highly correlated ($r^2 = 0.97$) and the slope of the
relationship is close to one (b = 1.23). In the Tachinidae 40
species were examined and again probability of discovering hosts
was correlated with ovariole number per ovary ($r^2 = 0.66$). Conopidae
fit into the general pattern seen in the Tachinidae.

ACKNOWLEDGEMENTS

I am grateful to D. M. Wood of the Biosystematics Research
Institute, Ottawa, for much help with the Tachinidae. I thank him,
C. Bouton and J. Thompson of the University of Illinois for review-
ing an earlier draft of this paper. Financial support during
this study was provided through the U. S. Public Health Training
Grant PH GM 1076.

LITERATURE CITED

Askew, R. R. 1971. Parasitic insects. American Elsevier, New
    York. 316 p.
Beaver, R. A. 1966. The development and expression of population
    tables for the bark beetle *Scolytus scolytus* (F.). J. Anim.
    Ecol. 35:27-41.
Berryman, A. A. 1973. Population dynamics of the fir engraver,
    Scolytus ventralis (Coleoptera: Scolytidae). I. Analysis
    of population behavior and survival from 1964-1971. Can.
    Entomol. 105:1465-1488.
Bess, H. A. 1936. The biology of *Leschenaultia exul* Townsend, a
    tachinid parasite of *Malacosoma americana* Fabricius and
    Malacosoma distria Hubner. Ann. Entomol. Soc. Amer. 29:593-613.
Blunck, H. 1944. Zur Kenntnis der Hyperparasiten von *Pieris
    brassicae* L. 1. Beitrag: *Mesochorus pectoralis* Ratz. und
    seine Bedeutung für den Massenwechsel des Kohlweisslings.
    Z. Angew. Entomol. 30:418-491.
Blunck, H. 1951. Zur Kenntnis der Hyperparasiten von *Pieris
    brassicae* L. 3. Beitrage: *Hemiteles simillimus* Taschb. nov.
    var. *sulcatus*. Kennzeichen und Verhalten der Vollkerfe. Z.
    Angew. Entomol. 32:335-405.
Bobb, M. L. 1964-65. The biology of *Mastrus argeae* (Viereck)
    (Hymenoptera: Ichneumonidae), a parasite of pine sawfly
    prepupae. Bull. Brooklyn Entomol. Soc. 59 and 60:53-62.
Bracken, G. K. 1966. Role of ten dietary vitamins on fecundity
    of the parasitoid *Exeristes comstockii* (Cresson) (Hymenoptera:
    Ichneumonidae). Can. Entomol. 98:918-922.
Clausen, C. P. 1940. Entomophagous insects. McGraw-Hill, New
    York. 688 p. Reprinted by Hafner, New York. 1962.
Clausen, C. P., J. L. King, and C. Teranishi. 1927. The parasites
    of *Popillia japonica* in Japan and Chosen (Korea) and their
    introduction into the United States. U. S. Dep. Agric. Bull.
    1429. 55 p.
Cleare, L. D. 1939. The Amazon fly (*Metagonistylum minense*,
    Towns.) in British Guiana. Bull. Entomol. Res. 30:85-102.
Dasch, C. E. 1964. Ichneumon-flies of America north of Mexico:
    5. Subfamily Diplazontinae. Mem. Amer. Entomol. Inst. 3.
    304 p.
Dowden, P. B. 1933. *Lydella nigripes* and *L. piniariae*, fly

parasites of certain tree-defoliating caterpillars. J. Agr. Res. 46:963-995.

Dowden, P. B. 1934. *Zenillia libatrix* Panzer, a tachinid parasite of the gypsy moth and the brown-tail moth. J. Agr. Res. 48: 97-114.

Dufour, L. 1851. Recherches anatomiques et physiologiques sur les Diptères. Mem. Acad. Sci., Inst. Nat. France., Sci. Math. Phys. 11:171-360.

Embree, D. G. 1965. The population dynamics of the winter moth in Nova Scotia, 1954-1962. Mem. Entomol. Soc. Can. 46:1-57.

Graham, A. R. 1953. Biology and establishment in Canada of *Mesoleius tenthredinis* Morley (Hymenoptera: Ichneumonidae), a parasite of the larch sawfly, *Pristiphora erichsonii* (Hartig) (Hymenoptera: Tenthredinidae). Quebec Soc. Prot. Plants Annu. Rep. 35:61-75.

Griffiths, K. J. 1969. The importance of coincidence in the functional and numerical responses of two parasites of the European pine sawfly *Neodiprion sertifer*. Can. Entomol. 101: 673-713.

Harcourt, D. G. 1963. Major mortality factors in the population dynamics of the diamond back moth, *Plutella maculipennis* (Curt). Mem. Entomol. Soc. Can. 32:55-66.

Harcourt, D. G. 1966. Major factors in survival of the immature stages of *Pieris rapae* (L.). Can. Entomol. 98:653-662.

Iwata, K. 1960. The comparative anatomy of the ovary in Hymenoptera. Part V. Ichneumonidae. Acta Hymenopterologica 1: 115-169.

Jaynes, H. A. 1933. The parasites of the surgarcane borer in Argentina and Peru, and their introduction into the United States. U. S. Dep. Agric. Tech. Bull. 363. 26 p.

Kerrich, G. J. 1942. Second review of literature concerning British Ichneumonidae (Hym.), with notes on palaearctic species. Trans. Soc. Brit. Entomol. 8:43-77.

Kerrich, G. J. 1969. Description of an ichneumonid (Hym.) that preys on egg-masses of weevils harmful to tea culture in Kenya. Bull. Entomol. Res. 59:469-472.

Klomp, H. 1966. The dynamics of a field population of the pine looper, *Bupalus piniarius* L. Adv. Ecol. Res. 3:207-305.

Kugler, J., and Z. Wollberg. 1967. Biology of *Agrothereutes tunetanus* Haber. (Hym. Ichneumonidae) an ectoparasite of *Orgyia dubia* Tausch. (Lep. Lymantriidae). Entomophaga 12: 363-379.

Landis, B. J. 1940. *Paradexodes epilachnae*, a tachinid parasite of the Mexican bean beetle. U. S. Dep. Agric. Tech. Bull. 721. 31 p.

Leius, K. 1961. Influence of food on fecundity and longevity of adults of *Itoplectis conquisitor* (Say) (Hymenoptera: Ichneumonidae). Can. Entomol. 93:771-800.

LeRoux, E. J., R. O. Paradis, and M. Hudon. 1963. Major mortality factors in the population dynamics of the eye-spotted bud moth, the pistol casebearer, the fruit-tree leaf roller, and the European corn borer in Quebec. Mem. Entomol. Soc. Can. 32:67-82.

McLeod, J. M. 1972. The Swaine jack pine sawfly, *Neodiprion swainei* life system: Evaluating the long-term effects of insecticide applications in Quebec. Envir. Entomol. 1:371-381.

Meijere, F. C. H. de. 1904. Beiträge zur Kenntnis der Biologie und der systematischen Verwandschaft der Conopiden. Tijdschr. Entomol. 46:144-225.

Miller, W. E. 1967. The European pine shoot moth - Ecology and control in the Lake States. For. Sci. Monogr. 14:1-72.

Morris, R. F., and C. A. Miller. 1954. The development of life tables for the spruce budworm. Can. J. Zool. 32:283-301.

Myers, J. G. 1934. The discovery and introduction of the Amazon fly. A new parasite for cane-borers (*Diatraea* spp.). Trop. Agr. (Trinidad) 11:191-195.

Pampel, W. 1914. Die weiblischen Geschlechtosorgane der Ichneumoniden. Z. Wiss. Zool. 108:290-357.

Pantel, J. 1910. Recherches sur les Diptères à larves entomobies. I. Caractères parasitiques aux points de vue biologique, éthologique et histologique. Cellule 26:25-216.

Parnell, J. R. 1966. Observations on the population fluctuations and life histories of the beetles *Bruchidius ater* (Bruchidae) and *Apion fuscirostre* (Curculionidae) on broom (*Sarothamnus scoparius*). J. Anim. Ecol. 35:157-188.

Pottinger, P. R., and E. J. LeRoux. 1971. The biology and dynamics of *Lithocolletis blancardella* (Lepidoptera: Gracillariidae) on apple in Quebec. Mem. Entomol. Soc. Canada 77:1-437.

Price, P. W. 1973a. Reproductive strategies in parasitoid wasps. Amer. Natur. 107:684-693.

Price, P. W. 1973b. Parasitoid strategies and community organization. Environ. Entomol. 2:623-626.

Price, P. W. 1974. Strategies for egg production. Evolution 28: 76-84.

Price, P. W. 1975a. Energy allocation in ephemeral adult insects. Ohio J. Sci. In press.

Price, P. W. 1975b. Insect Ecology. Wiley Interscience, New York. In press.

Réaumur, R. A. F. de. 1738. Mémoires pour servir à l'histoire des Insectes. Vol. 4. De l'Imprimerie Royale. Paris. 636 p.

Richards, O. W., and N. Waloff. 1961. A study of a natural population of *Phytodecta olivacea* (Forster) (Coleoptera: Chrysomelidae). Phil. Trans. Roy. Soc. London. B. 244:205-257.

Sabrosky, C. W., and P. H. Arnaud, Jr. 1965. Family Tachinidae. p. 961-1108. *In* A. Stone, C. W. Sabrosky, W. W. Wirth, R. H. Foote, and J. R. Coulson (eds.). A catalog of the Diptera of America north of Mexico. U. S. Dep. Agric., Agric. Res. Serv., Agr. Handbook 276.

Salt, G. 1931. Parasites of the wheat-stem sawfly, *Cephus pygmaeus* Linnaeus, in England. Bull. Entomol. Res. 22:479-545.

Samarasinghe, S., and E. J. LeRoux. 1966. The biology and dynamics of the oyster-shell scale, *Lepidosaphes ulmi* (L.) on apple in Quebec. Ann. Entomol. Soc. Quebec 11:206-292.

Stark, R. W. 1959. Population dynamics of the lodgepole needle miner, *Recurvaria starki* Freeman, in Canadian Rocky Mountain parks. Can. J. Zool. 37:917-943.

Thompson, W. R. 1928. A contribution to the study of dipteran parasites of the European earwig (*Forficula auricularia* L.). Parasitology 20:123-158.

Thompson, W. R. 1954. *Hyalomyodes triangulifera* Loew (Diptera: Tachinidae). Can. Entomol. 86:137-144.

Townes, H. 1969a. The genera of Ichneumonidae, Part 1. Mem. Amer. Entomol. Inst. 11. 300 p.

Townes, H. 1969b. The genera of Ichneumonidae, Part 2. Mem. Amer. Entomol. Inst. 12. 537 p.

Townes, H. 1969c. The genera of Ichneumonidae, Part 3. Mem. Amer. Entomol. Inst. 13. 307 p.

Townes, H. 1971. The genera of Ichneumonidae, Part 4. Mem. Amer. Entomol. Inst. 17. 372 p.

Townes, H., and M. Townes. 1951. Family Ichneumonidae. p. 184-409. *In* C. F. W. Muesebeck and K. V. Krombein (eds.). Hymenoptera of America north of Mexico. Synoptic catalog. U. S. Dep. Agric., Agric. Monogr. 2.

Townsend, C. H. T. 1935-1941. Manual of Myiology. Townsend and Filhos, Itaquaquecatuba, São Paulo, Brazil. Part 2 (1935), Part 3 (1936a), Part 4 (1936b), Part 7 (1938), Part 8 (1939), Part 10 (1940), Part 11 (1941).

Tripp, H. A. 1960. *Spathimeigenia spinigera* Townsend (Diptera: Tachinidae), a parasite of *Neodiprion swainei* Middleton (Hymenoptera: Tenthredinidae). Can. Entomol. 92:347-359.

Tripp, H. A. 1961. The biology of a hyperparasite, *Euceros frigidus* Cress. (Ichneumonidae) and description of the planidial stage. Can. Entomol. 93:40-58.

Tripp, H. A. 1962. The biology of *Perilampus hyalinus* Say (Hymenoptera: Perilampidae), a primary parasite of *Neodiprion swainei* Midd. (Hymenoptera: Diprionidae) in Quebec, with descriptions of the egg and larval stages. Can. Entomol. 94:1250-1270.

Waloff, N. 1968. Studies on the insect fauna on scotch broom *Sarothamnus scoparius* (L.) Wimmer. Adv. Ecol. Res. 5:87-208.

Wishart, G. 1945. *Aplomya caesar* (Aldrich), a tachinid parasite of the European corn borer. Can. Entomol. 77:157-167.

APPENDIX 1

Sources for life tables of natural populations of insect herbivores which provide survivorship data used in construction of Fig. 1 (except those in Group C).

Host Type and Insect Species	Habit	Reference

Type A - 70% or more of mortality by mid larval stage.

1 *Choristoneura fumiferana*	in bud or web	Morris and Miller 1954
2 *Bupalus piniarius*	on foliage	Klomp 1966
3 *Plutella maculipennis*	on foliage	Harcourt 1963
4 *Spilonota ocellana*	in bud	LeRoux et al. 1963
5 *Archips argyrospilus*	in leaf roll	LeRoux et al. 1963
6 *Rhyacionia buoliana*	in bud and shoot	Miller 1967
7 *Operophtera brumata*	on foliage	Embree 1965
8 *Recurvaria starki*	in needle mine	Stark 1959
9 *Sitona regensteinensis*	on root nodules	Waloff 1968
10 *Lepidosaphes ulmi*	on branches	Samarasinghe and LeRoux 1966
11 *Arytaina genistae*	on foliage	Waloff 1968

Type B - 40% or less of mortality by mid larval stage.

1 *Ostrinia nubilalis*	in stems	LeRoux et al. 1963
2 *Coleophora serratella*	in case	LeRoux et al. 1963
3 *Lithocolletis blancardella*	in leaf mine	Pottinger and LeRoux 1971
4 *Pieris rapae*	on foliage (cultured environment)	Harcourt 1966
5 *Neodiprion swainei*	on foliage, colonial	McLeod 1972
6 *Scolytus scolytus*	under bark	Beaver 1966
7 *Scolytus ventralis*	in twig mines	Berryman 1973

Between A and B

Leucoptera spartifoliella	in twig mines	Waloff 1968

C   Probably in Group A but uncertain as no data is provided for mid larval stage.

1 *Bruchidius ater*	in pod	Parnell 1966
2 *Apion fuscirostre*	in pod	Parnell 1966
3 *Phytodecta olivacea*	on foliage	Richards and Waloff 1961

APPENDIX 2

Estimates of fecundity in 17 species of Tachinidae and 8 species of
Ichneumonidae in relation to the number of ovarioles per ovary (used
in Fig. 3). In the Tachinidae generic names follow Sabrosky and
Arnaud (1965); in the Ichneumonidae Townes and Townes (1951),
Townes (1969a,b) are used. Specific names are those of the author
cited. † indicates species which attack stage 11 hosts used to
calculate the typical fecundity of species attacking this stage.

Family and Species	No. ovarioles per ovary	Estimated fecundity	References
TACHINIDAE			
Leschenaultia exul	119	2,800	Bess 1936
Leschenaultia exul	252	5,500	Bess 1936
Leschenaultia exul	230	4,000	Bess 1936
Zenillia libatrix	61	820	Dowden 1934
Zenillia libatrix	107	2,439	Dowden 1934
Zenillia libatrix	90	1,820	Dowden 1934
Oswaldia epilachnae	8	94	Landis 1940
Oswaldia epilachnae	27	324	Landis 1940
Aplomya caesar	48	800	Wishart 1945
Aplomya caesar	62	1,200	Wishart 1945
Bigonicheta setipennis	40	320	Thompson 1928
Spathimeigenia spinigera	18	250	Tripp 1960
Lydella nigripes	11	132	Dowden 1933
Lydella nigripes	14	168	Dowden 1933
Euphasiopteryx australis	70	450	Townsend 1936a
Pararchytas decisus	70	7,500	Townsend 1939
Bonnetia analis	45	2,000	Townsend 1939
Gymnocheta analis	30	550	Townsend 1939
Plagiomima similis	10	400	Townsend 1939
Cryptomeigenia prisca	12	75	Townsend 1940
Cryptomeigenia illionisensis	5	40	Townsend 1940
Cryptomeigenia eumythyroides	7	50	Townsend 1940
Chrysoexorista borealis	24	1,000	Townsend 1940
Eucelatoria armigera	3.5	150	Townsend 1940
ICHNEUMONIDAE			
Euceros frigidus	50	1,000	Tripp 1961
Exenterus canadensis	5	61	Price 1974, Griffiths 1969

Appendix 2, continued.

Family and Species	No. ovarioles per ovary	Estimated fecundity	References
Mastrus aciculatus †	4	7	Price 1974, Bobb 1964-5
Pleolophus indistinctus †	4	32	Price 1974, Price 1975a
Pleolophus basizonus †	4	42	Price 1974, Griffiths 1969
Hemiteles simillimus †	6	36	Blunck 1951
Mesochorus pectoralis †	11	100	Blunck 1944
Agrothereutes tunetanus	9	308	Kugler and Wollberg 1967

APPENDIX 3

Ichneumonidae of known attack characteristics, in order of species in Fig. 5, top to bottom, left to right. Generic names follow Townes (1969a,b,c, 1971).

Species	Oviposition type	Number ovarioles per ovary	References
Euceros pruinosus	on foliage	231	Iwata 1960
Euceros frigidus	" "	50	Tripp 1961
Pion fortipes	egg	30	Pampel 1914
Sympherta nipponica	"	25	Iwata 1960
Sympherta antilope	"	20	Iwata 1960
Collyria calcitrator	"	10	Pampel 1914, Salt 1931
Agasthenes swezeyi	"	4	Townes 1969b
Arachnoleter stagnalis	"	4	Kerrich 1942
Encrateola kerichoensis	"	4	Kerrich 1969
Gelis lymensis	"	3	Townes and Townes 1951
Gelis meabilis	"	3	Townes and Townes 1951
Gelis prosthesimae	"	3	Townes and Townes 1951
Olesicampe lophyri	very early larva	35	Price 1974
Syrphophilus bizonarius	" " "	28	Iwata 1960
Exetastes longipes	" " "	28	Iwata 1960
Exetastes fukuchizamensis	" " "	27	Iwata 1960

Appendix 3, continued.

Species	Oviposition type	Number ovarioles per ovary	References
Enicospilus americanus	very early larva	24	Price, unpublished
Diplazon tetragonus	" " "	19	Iwata 1960, Dasch 1964
Exetastes ishikawensis	" " "	18	Iwata 1960
Enizemum giganteum	" " "	17	Iwata 1960
Diplazon laetatorius	" " "	16	Iwata 1960, Dasch 1964
Exetastes sapporensis	" " "	14	Iwata 1960
Mesoleius tenthredinis	young larva	20	Graham 1953
Lissonota sapporensis	" "	19	Iwata 1960
Syzeuctus takaozanus	" "	15	Iwata 1960
Alloplasta longipetiolaris	" "	14	Iwata 1960
Apophua sapporensis	" "	13	Iwata 1960
Apophua tobensis	" "	12	Iwata 1960
Glypta glypta	" "	12	Iwata 1960
Syzeuctus sambonis	" "	12	Iwata 1960
Teleutaea japonica	" "	11	Iwata 1960
Teleutaea brischkei	" "	6	Iwata 1960
Ophion luteus	middle larva	15	Pampel 1914
Dictyonotus purpurascens	" "	14	Iwata 1960
Enicospilus tristrigatus	" "	13	Iwata 1960
Enicospilus sp.	" "	12	Iwata 1960
Ophion obscuratus	" "	11	Iwata 1960
Ophion takaozanus	" "	10	Iwata 1960
Enicospilus ramidulus	" "	8	Iwata 1960
Enicospilus fuscomaculatus	" "	6	Iwata 1960
Lamachus lophyri	old larva	10	Price 1974
Exenterus amictorius	" "	5	Price 1974
Exenterus diprionis	" "	5	Price 1974
Netelia testacea	" "	5	Pampel 1914
Netelia gracilipes	" "	5	Pampel 1914
Tryphon rutilator	" "	4	Pampel 1914
Tryphon trochanterus	" "	4	Pampel 1914
Polyblastus cothurnatus	" "	4	Pampel 1914
Dyspetes paerogaster	" "	3	Pampel 1914
Agrothereutes tunetanus	pupa, cocoon, or puparium	9	Kugler and Wollberg 1967
Pterocormus primatorius	" "	4	Pampel 1914
Stenichneumon culpator	" "	4	Pampel 1914
Stenichneumon pistatorius	" "	4	Pampel 1914
Protichneumon fusorius	" "	4	Pampel 1914
Mastrus aciculatus	" "	4	Price 1974

Appendix 3, continued.

Species	Oviposition type	Number ovarioles per ovary	References
Pleolophus indistinctus	pupa, cocoon, or puparium	4	Price 1974
Pleolophus basizonus	" "	4	Price 1974
Endasys subclavatus	" "	3	Price 1974
Gelis urbanus	" "	3	Price 1974

APPENDIX 4

Species of Tachinidae and Conopidae and their ovariole numbers per ovary of known attack characteristics, used in construction of Fig. 7. Generic names follow Sabrosky and Arnaud (1965), specific names the author cited; * indicates that genus is not given in Sabrosky and Arnaud (1965) as far as I can determine. In order from top to bottom, left to right in Fig. 7.

Family and Species	Placement of progeny	Number ovarioles per ovary	References
TACHINIDAE			
Leschenaultia exul	on food, eggs eaten	230	Bess 1936
Zenillia libatrix	" " "	85	Dowden 1934
Gonia atra	" " "	80	Pantel 1910
Aplomya caesar	" " "	55	Wishart 1945
Echinomyia fera *	on food or soil, planidial	140	Pantel 1910
Linnaemya vulpina	" "	100	Pantel 1910
Bonnetia compta	" "	90	Pantel 1910
Euphasiopteryx australis	" "	70	Townsend 1936a
Bigonicheta setipennis	" "	40	Pantel 1910
Metagonistylum minense[a]	at entrance to hole, planidial	34	Cleare 1939, Myers 1934
Paratheresia claripalpis[a]	" "	32	Jaynes 1933
Voria elata	on host larva (non-piercing ovipositor)	54	Pantel 1910
Carcelia cheloniae	" "	35	Pantel 1910
Tricholyga major *	" "	30	Pantel 1910
Phryxe vulgaris	" "	22	Pantel 1910
Oswaldia epilachnae	" "	18	Landis 1940

Appendix 4, continued.

Family and Species	Placement of progeny	Number ovarioles per ovary	References
Pseudomyothyria ancilla	on host larva (non-piercing ovipositor)	9	Townsend 1941
Spathimeigenia spinigera	in host larva (piercing ovipositor)	18	Tripp 1960
Compsilura concinnata	" "	15	Pantel 1910
Lydella piniariae	" "	13	Dowden 1933
Lydella nigripes	" "	13	Dowden 1933
Eucelatoria armigera	" "	4	Townsend 1940
Xanthomelanodes atripennis	on host adult (non-piercing ovipositor)	15	Townsend 1938
Gymnoclytia occidua	" "	13	Townsend 1938
Torynotachina quinteri †	" "	12	Townsend 1940
Gymnosoma rotundatum	" "	10	Pantel 1910, Dufour 1851
Hyperecteina cinerea	" "	10	Clausen et al. 1927
Xanthomelanodes peruana	" "	10	Townsend 1938
Clistomorpha didyma	" "	8	Townsend 1938
Hyalomyodes triangulifera	" "	6	Thompson 1954 Townsend 1938

CONOPIDAE

Conops rufipes	" "	17	Meijere 1904
Conops rufipes	" "	12	Dufour 1851
Myopa ferruginea	" "	12	Dufour 1851

TACHINIDAE

Epigrimyia polita	in host adult (piercing ovipositor)	15	Townsend 1938
Hemyda aurata	" "	14	Townsend 1938
Neocyptera dosiades	" "	12	Townsend 1938
Apinops ater	" "	12	Townsend 1938
Paraphorantha grandis	" "	12	Townsend 1938
Cylindromyia argentea	" "	11	Townsend 1938
Phasiomyia splendida	" "	9	Townsend 1938
Alophorella fenestrata	" "	8	Townsend 1938
Alophorella diversa	" "	8	Townsend 1938
Alophorella phasioides	" "	7	Townsend 1938

Appendix 4, continued.

[†]Townsend thought that this species had a piercing ovipositor.
However it is actually used as an elytra lifter and probably acts
in the same way as in *Velocia* spp. in which an unincubated egg is
deposited on the inner surface of the elytra of chrysomelid beetles
(D. M. Wood, personal communication).

[a]Ovariole number of species not provided in source.   This was
estimated from the known fecundity using Fig. 3.

SUCCESSION OF r AND K STRATEGISTS IN PARASITOIDS

Don C. Force

Department of Biological Science, California State

Polytechnic University, Pomona, California 91768, U.S.A.

THEORY OF r AND K SELECTION

The concept of r and K strategems and their development during the evolution of organisms is certainly not new, although these particular terms to describe the process have been used for only a short period of time. As a matter of fact, the concept is implicit in some of the writings of Charles Darwin, among others. But, however old the basic theory is, there is presently a great deal of uncertainty as to how important these stratagems are in the functioning of organisms within ecological communities. There have also been questions as to whether or not genetic selection can proceed in the manner prescribed by the concept; even the appropriateness of the symbols "r" and "K" have been criticized. And to add to the dilemma, I suppose it is not altogether suitable to use the term "stratagem" when discussing the r and K theory since the dictionary defines stratagem as an artifice--a trick or way to deceive the enemy--and certainly this is a questionable manner in which to describe the process of evolution or development at any level of an ecosystem.

Apparently MacArthur and Wilson (1967) originated the terms r selection and K selection. Pianka (1970) has commented on the fact that perhaps the terms are unfortunate since they "... invoke the much overused logistic equation." However, the concept appears clear enough with its two opposing kinds of selective forces operating. Dobzhansky (1950) very aptly defined the conditions for the process. He described the differences between temperate or arctic environments and those of tropical regions. The former are characterized by great physical perturbations which must have a profound effect on the selection of the organisms living there.

These organisms are selected for great physiological tolerances and other mechanisms that enable them to cope with the physical factors; and since populations are often decimated in these climates, rapid replacement is particularly important for them. In tropical environments, on the other hand, where physical problems are seldom encountered, organisms are more constantly crowded and selection is influenced more by biotic factors. Competition then becomes a primary selective force because the populations are more constantly at or near the carrying capacity of the environment.

MacArthur and Wilson (1967) described the process in more theoretical terms. They set the stage by postulating two possible situations: (1) an environment in which a population is expanding, and (2) an environment in which a population is crowded. They assumed relatively stable conditions in both cases, i.e., no great physical changes occurring to radically change the environments. They then envisioned the first situation occurring with no crowding and plentiful food. Under these conditions, organisms should evolve toward productivity and large families since there is no shortage of resources. Wastage of resources can be permitted in this case. This kind of evolution is called r selection after the parameter r or r_{max} in the Verhulst-Pearl logistic equation. On the other hand, under conditions of crowding, selection should evolve toward feeding efficiency since resources are in short supply. The organism that can replace itself with the least waste of resources is the most fit in this environment. Evolution here is called K selection after the parameter K or environmental carrying capacity in the equation. MacArthur and Wilson considered briefly the implications of the theory biogeographically, but put particular emphasis on its possible effect in the case of island populations. Other authors have re-stated the theory in terms similar to those of MacArthur and Wilson. Gadgil and Bossert (1970), for example, also concluded that the conditions of resources in an environment are responsible for the kind of selection that takes place; r selection occurring where resources are plentiful and K selection where they are not. MacArthur (1972) reminded us that there are many other selective mechanisms which are possibly operating besides r and K selection, and that it may be by no means the primary evolutionary mechanism in most communities, nor even a very important one.

Pianka (1970) thoroughly treated the subject and its history, and listed some of the correlates of both r and K selection. He proposed that terrestrial vertebrates and perennial plants are pre-dominately K-selected, whereas terrestrial invertebrates and annual plants are more r-selected. But he cautioned against rigidly describing any organisms as being either completely r- or K-selected, and indicated rather that a compromise is generally necessary. Pianka also visualized an r-K continuum with organisms positioned along it according to the relative differences in their evolved

characteristics. However, he clearly pointed out that in the
real world, selective forces are constantly changing in form or in
magnitude; therefore, the organisms in a population must also be
constantly changing physiologically and genetically with these
stresses. Hence, the process is a dynamic phenomenon with organisms
continually shifting, however slightly, toward either the r or the K
end of the continuum.

Gadgil and Solbrig (1972) also discussed r and K selection in
detail, and attempted to clarify certain aspects of the theory. In
fact, nearly everyone who has treated the theory has "adjusted" it
more or less in the process. Perhaps this is inevitable when debat-
able concepts are being mulled over. These authors objected to the
assumption that increased or decreased birth rate by itself is evidence of increased r or K selection respectively. For this reason
they objected to the use by MacArthur and Wilson (1967) of the evidence presented by Cody (1966) that greater clutch size in bird
species from harsher environments indicates r selection. Gadgil
and Solbrig maintained that it is necessary, when providing evidence
for r selection, to show that reproductive output uses a greater
relative amount of the total resources available, and not merely
that reproduction has increased. This argument would appear to
have some validity. They also objected to necessarily equating
r strategists with colonizing or fugitive species, a correlation
that might be taken more or less for granted by many ecologists.
Their objection was that colonizers must be good dispersers, but
that there are situations involving high density independent mortality (and therefore clearly r strategist environments) where high
dispersal capability is not a necessity.

Hairston et al. (1970) have criticized the r and K theory from
the standpoint that selection cannot act directly upon the innate
capacity for increase (r), but rather upon the birth or death rate
per individual. Therefore, they proposed replacing the terms r
and K selection with b (birth) and d (death) selection. Pianka
(1972) apparently reconciled the problem and further clarified
the entire concept. King and Anderson (1971) have considered the
genetic aspects of r and K selection in relation to constant and
variable environmental models. They found that K selection always
predominates in a constant environment, whereas either r or K
selection may predominate under variable conditions. Such studies
as this should lead to greater refinement of the theory.

Various authors (Pianka 1970, Gadgil and Solbrig 1972, Force
1972 and others) have described the characteristics associated
with both r and K strategists. Table 1 lists certain of these;
however, one must remember that these characteristics can be used
only as relative descriptions. Indeed, perhaps the most necessary
condition involving the entire r and K concept is that it must
always be applied relatively; i.e., the attributes of a certain

Table 1. Certain characteristics of *r* and *K* strategists.

r strategists	*K* strategists
Found most abundantly where density independent factors predominate:	Found most abundantly where density dependent factors predominate:
1. temperate and arctic climates 2. edges of population range 3. disturbed situations	1. tropical climate 2. center of population range 3. undisturbed situations
Tend to have a high fecundity (perhaps small or undeveloped progeny) and short generation time (high *r*)	Tend to have a low fecundity (perhaps large or well-developed progeny) and long generation time (low *r*)
Usually reproduce early and rapidly, then die early	Usually delay reproduction, then reproduce slowly and live a long time.
Tend to be small in size	Tend to be large in size
Dependent upon abundant food resources; may be wasteful of resources; allocate relatively more of resources for reproduction	Can exist with limited food resources; highly competitive for resources; allocate relatively more resources to competitive strategies
Poor competitors--opportunists	Good competitors--specialists
May have greater tolerances to harsh environmental conditions	May have restricted tolerances to harsh environmental conditions

organism can be declared either those of an r strategist or a K strategist only in comparison with another organism or group of organisms. Thus, organism A may be an r strategist compared to organism B, but a K strategist when compared to organism C.

STUDIES REVEALING r AND K SELECTION

Since the concept of r and K selection is rather new--at least as a formalized theory--very few studies have been performed to test its veracity. It is not an easy task to gather the data necessary to indicate r or K selection. However, Solbrig (1971) and Gadgil and Solbrig (1972) have produced evidence that the process occurs in wild flowers. These authors have shown that among dandelions (*Taraxacum officinale* L.), which are r strategists in relation to most other plants, there are different biotypes which have become more r- or K-selected according to their particular habitat. Thus, it was demonstrated that those plants occupying a very disturbed habitat were more r-selected than were those growing under less disturbed conditions. The more r-selected plants produced more and smaller seeds, and produced them earlier. The K-selected plants produced more vegetation and outcompeted the others in growth. These authors also studied various species of goldenrod (*Solidago*) and found the most r-selected species were most common in the more harsh, unstable areas as would be expected from the theory.

Huey et al. (1974), Pianka (1970, 1972), Tinkle (1969), and Tinkle et al. (1970) have gathered considerable evidence that indicates strong r and K selection among lizards. Lizards are ideal organisms to use in the study of r and K selection because their clutches or litters can be easily measured for both numbers and weight, and they seldom exert parental care. Therefore, the amount of energy going into reproductive effort can be readily ascertained. In addition, their reproductive strategies are rather variable. For example, some are viviparous, others are oviparous; some mature the first year, others take longer than a year to mature; clutch size varies from one to as many as 30, and the frequency of clutch production varies considerably. Among mammals, Fleming (1974) has indicated the possible occurrence of r and K selection in tropical rodents.

As far as insects are concerned, the data on r and K selection are just as scarce as for other kinds of organisms. Unfortunately, most studies that might have provided the kind of data needed, fall short by various degrees. For example, many studies dealing with competition among various insects have been performed solely under laboratory conditions. Laboratory tests by themselves can often tell us a great deal about the competitive characteristics and the capacity for increase of an organism, and therefore, whether it

should be more *r*- or *K*-selected than another when comparisons are
made; but these studies cannot tell us whether the organisms act
like *r* or *K* strategists in nature. And field studies by themselves
are meaningless without adequate laboratory data on innate capacities for increase (fecundity counts are not always enough) and
competitive attributes. The situation is just as dismal for
parasitoids as for other kinds of insects. The reader might assume
that biological control studies could help since so many species
of parasitoids have been investigated in this field of endeavor.
Price (1973b) commented on this possibility. He said,

> "It would be interesting to survey the qualities of species
> that have been successfully introduced as influential
> biological control agents in relation to their status as
> *r* or *K* strategists. No doubt the effort would be frustrated
> either by lack of biological information on the species
> concerned or by lack of data on their impact on host and
> sympatric parasitoid populations."

In fact, adequate data are exceedingly scarce. But before discussing some of those data that are available, let me speculate briefly
on how it might be possible for *r* and *K* strategist parasitoids to
take their places in a community.

DEVELOPMENT OF *r* AND *K* STRATEGISTS IN COMMUNITIES

Goulden (1969), Slobodkin and Sanders (1969) and others have
suggested possible mechanisms whereby both *r* and *K* strategists
could be incorporated into community succession. Goulden, for
example, postulated that immature or disturbed areas are characterized by having few species, one of which is very likely dominant.
These early dominants are sometimes called opportunists and may
be quite tolerant of environmental perturbations. As the physical
environment becomes more stable because of ecological succession,
less tolerant but more specialized species, which are better
competitors invade the community. Now, if it is reasonable to
expect this kind of process to occur, it would probably begin in
areas with unstable conditions and therefore high density independent mortality. Gadgil and Solbrig (1972) have identified three
types of situations where density independent mortality should be
high: (1) environments which are permanent in both space and time
with continuously occurring factors causing high density independent mortality; (2) environments permanent in space but with
discontinuously occurring factors causing high density independent
mortality periodically (e.g., an adverse season during the year);
(3) environments temporary in time and space such as disturbed
areas where density independent mortality is high during the periods
of disturbance (e.g., any of a variety of agricultural areas).
If an insect host were able to adapt to one of these kinds of areas,

any parasitoid successfully attacking that host insect would have to possess one or more of the following attributes: (1) great physiological tolerances to withstand regular or irregular physical perturbations; (2) great dispersal capabilities so that new host populations could be located quickly in case of local disaster; (3) a high potential rate of numerical increase so that normal population numbers could be established quickly between perturbations or disturbances. These are characteristics of r strategists, and so the early dominants or opportunists would be relatively r selected.

As the ecosystem develops, more physical stability is likely to occur, allowing the establishment of parasitoid species which cannot tolerate great physical stresses. These species, in order to become established, must have characteristics which are more K-selected; i.e., they must be able to usurp at least part of the niche of the r strategist through competitive mechanisms. This can be accomplished providing the invader is not behaviorally restrained in attacking hosts already parasitized by the r strategists, and that the progeny of the latter are killed in most cases of multiple parasitism. A series of invasions can be visualized in time, each new invasion accomplished by a parasitoid species which is more K-selected than one or more of the established species. The r strategists will survive if they are capable of maintaining for themselves any part of their original niche, and thereby more species will gradually accumulate in the system, and niches will become progressively smaller. Eventually the more K-selected organisms will dominate the community because of their competitive advantages. The r strategists may be relatively rare, but will possibly survive best either at the outer range limits of the host, where physical factors are more stringent, or under disturbed conditions anywhere within the community.

EVIDENCE OF r AND K STRATEGISTS AMONG PARASITOIDS

To my knowledge only two host-parasitoid communities have been analyzed as to the r and K stratagems utilized by the member parasitoids. One of these is the Swaine jack pine sawfly, *Neodiprion swainei* Middleton, and its parasitoid complex, which has been treated by Price in a number of important articles (e.g., Price 1970a, b, 1971, 1972a, b, 1973a, b, 1974, Price and Tripp 1972). Pertaining to r and K strategists in particular, this community was discussed by Price (1973b). The other community is the cecidomyiid midge, *Rhopalomyia californica* Felt, that forms galls on coyote brush (*Baccharis pilularis* De Candolle) on the California coast, and the parasitoids that attack it, reported by Force (1970, 1972, 1974). Price (1973b) analyzed the parasitoids of the jack pine sawfly from three different successional standpoints. He observed the characteristic changes that occur in the parasitoid complex because of

(1) temporal plant species displacement, (2) spatial variations in host-insect densities, and (3) temporal (seasonal) variations in host-insect availability for parasitization. Table 2 shows certain of his data. He found that as succession in plant cover occurs in a burned area being recolonized by jack pine, the habitat for cocoon (pupal) parasitoids of the sawfly (which pupates on the ground) changes from rather harsh to more favorable conditions. As the transition occurs, the number of parasitoid species grows progressively larger to a certain point in plant succession, and then declines slightly in the later stages of the process. This general trend in species numbers has been observed in other plant successions. Price also found a progressive change in kinds of sawfly parasitoids from the perimeter of an infestation, where sawflies are sporadic, to the center, where they are much more abundant. Larval parasitoids were more common at the physically more rigorous perimeter, but were gradually replaced by cocoon parasitoids approaching the center; and total numbers of parasitoid species increased correspondingly from edge to center. Finally, Price observed progressive differences in parasitoids from the more abundant larval stage of a host-insect generation, to the less abundant pupal stage. Again, numbers of parasitoid species increased from early in a generation to late, and the larval parasitoids were more general in their host-searching habits than were the pupal. The majority of pupal parasitoids practiced facultative hyperparasitism (attacking either the host insect or another parasitoid species), a mechanism which increases their competitive advantage because they readily destroy larval parasitoids already present in the host insect.

All of these data presented by Price give credence to the theory that r and K selection is strong and important in this group of insects. As the theory predicts, he found generally increasing numbers of parasitoid species as the various types of succession progressed, as well as differences in adaptations of the organisms. Larval parasitoids, which are predominant only in early or peripheral stages of succession, were found to be more r-selected; i.e., they have such traits as higher fecundities and more mobility in locating hosts. The pupal parasitoids have lower fecundities and better competitive abilities, hence they are more K-selected, and tend to displace the r strategists from the longer and better-established host populations.

The midge, *Rhopalomyia californica*, and its parasitoid complex were the object of my own investigations. I discovered from laboratory and glasshouse studies (Force 1970) that four of the parasitoids attacking this midge had reproductive capacities inversely related to their competitive abilities. It occurred to me that r and K selection was an evolutionary mechanism that would explain this finding, and also provide an explanation for the observation that endemic insect hosts very often are attacked by large numbers of parasitoid species, and hence interspecific competition among them

Table 2. Characteristic changes that occur in the parasitoid complex of *Neodiprion swainei* during three types of succession. Data from Price (1973b).

(a) Temporal plant species succession under pine cover

	(early succession)					(late succession)
	dry sand	pine litter	litter & lichen	lichen	lichen & moss	moss
No. of parasitoid species present	2	2	4	6	5	5

(b) Spatial host-insect density succession--transect from light to heavy density

	(light density)						(heavy density)
% parasitism by larval parasitoids	13	4	1	2	3	4	3
% parasitism by pupal parasitoids	0	1	7	2	5	2	9
No. of parasitoid species present	2	3	3	3	5	4	5

(c) Temporal host-insect succession--stage of generation

	early larva	late larva	eonymphal	pupal
No. of parasitoid species present	2	2	4	7

must be severe (Force 1972). Table 3 provides some of the evidence from which judgments were made concerning the relative r- and K-selected status of the six most common species of *Rhopalomyia* parasitoids. *Tetrastichus* is clearly the most r-selected because of its high r and inferior competitive characteristics. When parasitization by other species is high as it often is, *Tetrastichus* is disadvantaged because of its reluctance to attack already-parasitized hosts, and therefore this species is usually low in numbers. *Platygaster* is an egg parasitoid, so it always attacks the host prior to parasitization by the other species, all of which are larval parasitoids. However, the others (except *Tetrastichus*) freely attack hosts already parasitized by *Platygaster*, killing the latter in the process. *Torymus baccharicidis* larvae successfully compete with larvae of all other species except *T. koebelei*; however, the adult female of the former species never parasitizes a host already containing a larva of the latter, and so saves progeny that would otherwise be lost. Both *Amblymerus* and *Zatropis* are facultative hyperparasitoids, and so successfully attack either the midge or other parasitoids. *Zatropis* prefers to attack hosts already parasitized by *Tetrastichus*, thus further inhibiting the disadvantaged *Tetrastichus*. Data from field studies of the midge indicated that intense competition occurs among its parasitoids, and that the more productive species were suppressed by the less productive but more competitive species--a situation to be expected under r and K selection (Force 1974). One particular incident occurred during the course of the field studies that contributed convincing evidence of the interaction of r and K parasitoids in this community. One of the collecting sites was disturbed by the removal of many of the *Baccharis* plants and the cutting of the remaining plants close to the ground. At the time of the next collection several weeks later, the most r-selected parasitoid (*Tetrastichus*) had increased its degree of parasitization of the host population from 1 to 46% at the expense of the other, more K selected species. Gradually as the plants recovered over the next few months, the other parasitoid species returned to their normal status in the community, forcing *Tetrastichus* back to its average 1 to 3% parasitization.

It is most interesting to compare the *Neodiprion* and *Rhopalomyia* communities; not so much because of their similarities as one might assume, but rather because of their differences. Even though both are host-parasitoid systems, they are nonetheless vastly different in a number of ways. For example, they occur in completely different climates. *Neodiprion* inhabits a typical mid-latitude temperate climate with severe winters and warm, wet summers. *Rhopalomyia*, on the other hand, inhabits a Mediterranean type climate with warm winters and mild, moderately dry summers. The sawfly infests a tree and is unprotected in the egg and larval stages, then drops to the ground to pupate where it is better protected in the litter. The midge infests a bush and is unprotected in the egg stage, but then

Table 3.   Certain characteristics of six parasitoids that attack
           Rhopalomyia californica. Note that the species become
           progressively more *K* selected reading from the top
           to bottom.

Species	Characteristics
Tetrastichus sp.	r = 0.200. High restraint in parasitizing hosts already parasitized by other species. Larvae usually lose (die) in competition for host with larvae of other species.
Platygaster californica	r = 0.092. Larvae nearly always die in competition for host with larvae of other species.
Torymus koebelei	r = 0.090. No restraint in parasitizing hosts already parasitized by another species. Larvae nearly always win in competition for host with larvae of other species.
Torymus baccharicidis	r = 0.080. No restrain in parasitizing hosts parasitized by another species --except in case of *T. koebelei*. Larvae nearly always win in competition for host with larvae of other species.
Amblymerus sp.	r not determined. Successfully parasitizes larvae of either host insect (*Rhopalomyia*) or another parasitoid.
Zatropis sp.	r not determined. Same as *Amblymerus*, except prefers to parasitize larvae of *Tetrastichus* over other possible species.

resides inside a gall during the larval and pupal stages. Because
of these and other differences, one might expect entirely different
sorts of functional strategies to evolve within the community.
The fact that the sawfly and its parasitoids are univoltine, whereas
the midge and its parasitoids are multivoltine and continue to
reproduce throughout the year, indicates how great the differences
are in the two communities. Yet they appear to have developed and
to be functioning in a very similar manner when analyzed from the
r and K selection point of view.

We would really like to know a great deal more about the long-
term history of these two communities. For example, Price (1973b)
reported that there are 19 species of parasitoids that attack
Neodiprion. Two of these are exotic species that were imported
to control another pest insect, so there are 17 endemic species.
It would be enlightening to know whether the number has been
constantly increasing in the past; or perhaps there were more than
17 at one time, and the number has since decreased. Many endemic
host-insects have perhaps 30 or 40 species of parasitoids attacking
them. Does a greater number of parasitoids necessarily mean a longer
evolutionary history? In the case of *Rhopalomyia*, there are perhaps
a dozen species of parasitoids that attack it in the coastal area
of California. Only seven of these are fairly common. The others
are sporadic (even rare) and appear to parasitize *Rhopalomyia* more
by accident than anything else. Again, we would like to know how
long these parasitoids have been associated with the host insect,
and how long the host insect has been associated with the host plant,
Baccharis. Information such as this could elucidate many of the
problems we have in interpreting our findings.

In spite of the fact that the applied field of biological control does not appear to offer much in the way of incontestable
evidence regarding r and K strategists among parasitoids, there is
a scattering of tantalizing (but unfortunately often ambiguous)
data that may indicate the development of these strategies in other
host-parasitoid systems. With the vast amount of literature
accumulated in this field, it may seem strange that there is not
more evidence available. But again, most of these studies have been
rather superficial with respect to such things as the reproductive
rates of parasitoids in combination with interspecific competition
among them. And, of course, the systems studied in these endeavors
are generally not natural communities, but rather parts of communities that have been introduced into new areas. What effect this
disruption may have on community balance and organization, and on
the behavior and interactions of the organisms is a matter of
speculation.

One of the more complete and fascinating studies ever performed
on an introduced insect host-parasitoid system was undertaken by
DeBach and various of his associates on the California red scale,

Aonidiella aurantii (Maskell), and its parasitoids in southern
California. Different phases of this investigation were reported
in various places, but much of the data were incorporated into
an article by DeBach and Sundby (1963), which dealt primarily with
the theory of competitive displacement of ecological homologues.
Many of the data are inconclusive or difficult to interpret in
terms of r and K selection, but there are certain findings that
suggest intriguing possibilities from this standpoint.

The study included six species of parasitoids altogether:
four species of ectoparasitoids in the genus *Aphytis*, and two
species of endoparasitoids, *Comperiella bifasciata* and *Prospaltella
perniciosi*. All were imported into this country at various times
to control the California red scale, which was also (accidently)
introduced. The various species of *Aphytis* were found to be apparent
ecological homologues, and therefore certain species had displaced
others--or had not become established in one case--depending upon
slight differences in ecological preferences, fecundities, etc.
The endoparasitoids, on the other hand, were not displaced by the
ectoparasitoids ostensibly because the niches of the two groups
were not that similar. A great amount of effort went into studying
the fecundities of the four *Aphytis* species, both in single species
and mixed species culture; also, both larval and adult competitive
behavior was investigated. The results of these studies indicated
that the differences in fecundities among the various species were
primarily responsible for the displacement of one by another, at
least in the laboratory. Eventually the species with the higher
fecundity under a particular set of physical conditions completely
displaced another species with even a slightly lower fecundity.
A secondary and much less important factor in this displacement
was thought to be larval competition. At first I considered
the possibility of analyzing these data as possible evidence of
r and K selection among the various *Aphytis* species, but abandoned
the effort when I realized that the close relationships between
these insects prevent this kind of a comparison because their
characteristics are much too similar. Differences among the species
are often slight and difficult to measure accurately, and in any
case the data produced by DeBach and associates present some problems. For example, they used the oleander scale as host insect
material for the *Aphytis* tests, but discovered later that the
fecundities are different on these hosts than on California red
scale. Besides, if these species of *Aphytis* are indeed ecological
homologues, there is some question (theoretically at least) whether
it is possible for them to be differently r and K-selected, one with
another. However, a comparison between one or more *Aphytis* species
and the two endoparasitoids seemed to be a better possibility.

DeBach and Sundby had also studied the reproductive capacities
and certain aspects of the competitive characteristics of both
endoparasitoids, *Comperiella* and *Prospaltella*; and they compared

these data with those of one species of *Aphytis*. Table 4 shows
some of the findings from these studies. They estimated that after
a period of about 60 days at approximately 26°C, there would be a
vast difference in the number of female progeny produced by each
species. *Prospaltella* would be capable of producing nearly seven
times more than would *Aphytis*; and *Aphytis* over 21 times more than
Comperiella. Unfortunately, direct larval competition between the
species was not realistically studied, and this aspect of the situation could be very important to the entire picture. Often when a
host insect is multiparasitized by both an ecto- and endoparasitoid,
the ectoparasitoid kills the other, or in any case is the survivor.
However, certain other indications of interference were noted. They
found, for example, that *Prospaltella* was not able to persist in
laboratory colonies with the other two species. They found that
when small numbers of adult females of all three species were tested
for reproduction in single and mixed species cultures, both *Aphytis*
and *Comperiella* produced about the same number of progeny when
tested alone or with each other, but both produced more progeny in
combination with *Prospaltella*. On the other hand, *Prospaltella*
produced a much larger number of progeny when cultured alone than in
mixed culture with either of the others. Also, they found that
Aphytis feeds on and thereby kills small scale insects that
Prospaltella prefers to parasitize.

In the field, *Aphytis* became dominant in relation to both
endoparasitoids, but *Comperiella* appears to have done better than
might be expected with its low reproductive powers. It exists in
the inland, hotter areas which are optimal for *Aphytis*, and yet
Comperiella is apparently common and, "... sometimes more numerous
than the *Aphytis* species at certain times or places," according to
DeBach and Sundby (1963). *Prospaltella* was at one time more common
in the inland areas, but apparently has been crowded out by competition from the other species. In 1961 it occurred almost
exclusively in coastal areas where the climate appears to be too
cool for *Aphytis* or *Comperiella*. These findings tend to indicate
an inverse relationship between the reproductive capacities of
these three species and their competitive abilities--the same kind
of situation as was noted with the parasitoids of *Neodiprion*
and *Rhopalomyia*. *Prospaltella* has the greatest capacity for
numerical increase of the three species, but is thwarted in
realizing this potential advantage because of its poor competitive
characteristics. *Comperiella* has a much more limited capacity
for increase than *Aphytis*, yet it appears to do better in competition
with *Aphytis* under both laboratory and field conditions than its
reproductive potential would indicate possible, thereby manifesting
the possession of a competitive advantage of some sort.

Flanders (1971) has made this same host-parasitoid complex
the object of some candid observations, which tend to corroborate
the above evidence concerning *r* and *K* selection. He discussed

Table 4.   Certain statistics derived from the results of single
           species and mixed species rearing tests of three
           parasitoid species attacking *Aonidiella aurantii*.
           Data from DeBach and Sundby (1963).

	Parasitoid species		
	Aphytis	*Comperiella*	*Prospaltella*
Theoretical no. of females produced after about 60 days	14,641	676	97,336
Mean no. progeny/ female in single species tests	33.7	26.6	17.1
Mean no. progeny/ female in mixed species tests	29.8	27.2	
	57.5		5.3
		34.2	11.9

the fact that ectoparasitoids in general, and different species
of *Aphytis* in particular, are often considered to be the best
host regulative parasitoids of various scale insects. However,
he presented some rather convincing evidence that certain endoparasitoids might be more effective in controlling host populations
but have never been given a chance to prove their potential abilities in biological control programs because they have almost always
been introduced into areas with already existing populations of
ectoparasitoids, which tend to possess more effective competitive
characteristics. Concerning the south China race of *Prospaltella
perniciosa* (which is the one that was introduced into California),
he stated that often this species is not readily found parasitizing its host insect in its native habitat. When present, it is
most abundant in very sparse infestations of scale that *Aphytis*
apparently has trouble finding. This condition is what one might
expect of an r strategist in its native environment; i.e., (1)
restricted numbers because of competitive disadvantages with more
K-selected species, and (2) found most abundantly in marginal areas
of one type or another--or at least where the K strategists are
scarce or absent. In California also, *Prospaltella* is found only
in what probably should be considered marginal conditions--coastal

areas too cool for *Aphytis* and *Comperiella*.

Undoubtedly other scattered and probably controversial evidence exists in biological control and ecological literature concerning possible *r* and *K* strategies among parasitoids. Hopefully, there is enough interest in the subject that some of these data can be brought together and analyzed. More likely, however, if we desire to give the theory of *r* and *K* selection among parasitoids a thorough and realistic test, new and better data than available now will have to be provided.

SUMMARY

The basic concept of *r* and *K* selection is not new in evolutionary thinking. However, evidence indicating the possibility that certain organisms have evolved in this manner is just now beginning to be brought forth. There are a variety of questions to be asked about the mechanism by which *r* and *K* strategists might develop within a community or ecosystem. The most likely mechanism appears to be a gradual usurpation of the *r* strategist niches by various *K* strategists, thereby decreasing the size of niches and increasing the number of species in the community. Only two host-parasitoid communities have been analyzed from the standpoint of *r* and *K* selection; both possess members that fit the criteria of *r* and *K* strategists. Tantalizing clues from certain other host-parasitoid communities indicate that they too may have developed in a similar manner.

LITERATURE CITED

Cody, M. L. 1966. A general theory of clutch size. Evolution 20:174-184.

DeBach, P., and P. A. Sundby. 1963. Competitive displacement between ecological homologues. Hilgardia 34:105-166.

Dobzhansky, T. 1950. Evolution in the tropics. Amer. Sci. 38:209-221.

Flanders, S. E. 1971. Multiple parasitism of armored coccids (Homoptera) by host-regulative aphelinids (Hymenoptera); ectoparasites versus endoparasites. Can. Entomol. 103:857-872.

Fleming, T. H. 1974. The population ecology of two species of Costa Rican heteromyid rodents. Ecology 55:493-510.

Force, D. C. 1970. Competition among four hymenopterous parasites of an endemic insect host. Ann. Entomol. Soc. Amer. 63:1675-1688.

Force, D. C. 1972. *r-* and *K-* strategists in endemic host-parasitoid communities. Bull. Entomol. Soc. Amer. 18:135-137.

Force, D. C. 1974. Ecology of insect host-parasitoid communities. Science 184:624-632.

Gadgil, M., and W. H. Bossert. 1970. Life historical consequences of natural selection. Amer. Natur. 104:1-24.

Gadgil. M., and O. T. Solbrig. 1972. The concept of *r*- and *K*-selection evidence from wild flowers and some theoretical considerations. Amer. Natur. 106:14-31.

Goulden, C. E. 1969. Temporal changes in diversity, p. 96-102. In Diversity and stability in ecolgoical systems. Brookhaven Symposia in Biology, no. 22. Brookhaven Natl. Laboratory.

Hairston, N. G., D. W. Tinkle, and H. M. Wilbur. 1970. Natural selection and the parameters of population growth. J. Wildl. Manage. 34:681-690.

Huey, R. B., E. R. Pianka, M. E. Egan, and L. W. Coons. 1974. Ecological shifts in sympatry: Kalahari fossorial lizards (*Typhlosaurus*). Ecology 55:304-316.

King, C. E., and W. W. Anderson. 1971. Age-specific selection. II The interaction between *r* and *K* during population growth. Amer. Natur. 105:137-156.

MacArthur, R. H. 1972. Geographical ecology: patterns in the distribution of species. Harper and Row, New York. 269 pp.

MacArthur, R. H., and E. O. Wilson. 1967. The theory of island biogeography. Princeton Univ. Press. Princeton, N.J. 203 pp.

Pianka, E. R. 1970. On *r*- and *K*-selection. Amer. Natur. 104: 592-597.

Pianka, E. R. 1972. *r* and *K* selection or *b* and *d* selection? Amer. Natur. 106:581-588.

Price, P. W. 1970a. Characteristics permitting coexistence among parasitoids of a sawfly in Quebec. Ecology 51:445-454.

Price, P. W. 1970b. Trail odors: Recognition by insects parasitic on cocoons. Science 170:546-547.

Price, P. W. 1971. Niche breadth and dominance of parasitic insects sharing the same host species. Ecology 52:587-596.

Price, P. W. 1972a. Parasitoids utilizing the same host: Adaptive nature of differences in size and form. Ecology 53:190-195.

Price, P. W. 1972b. Behavior of the parasitoid *Pleolophus basizonus* (Hymenoptera: Ichneumonidae) in response to changes in host and parasitoid density. Can. Entomol. 104:129-140.

Price, P. W. 1973a. Reproductive strategies in parasitoid wasps. Amer. Natur. 107:684-693.

Price, P. W. 1973b. Parasitoid strategies and community organization. Envir. Entomol. 2:623-626.

Price, P. W. 1974. Strategies for egg production. Evolution 28: 76-84.

Price, P. W., and H. A. Tripp. 1972. Activity patterns of parasitoids on the Swaine jack pine sawfly, *Neodiprion swainei* (Hymenoptera: Diprionidae) and parasitoid impact on the host. Can. Entomol. 104:1003-1016.

Slobodkin, L. B., and H. L. Sanders. 1969. On the contribution of environmental predictability to species diversity, p. 82-95. In Diversity and stability in ecological systems. Brookhaven Symposia in Biology, no. 22. Brookhaven Natl. Laboratory.

Solbrig, O. T.   1971.   The population biology of dandelions.   Amer.
     Sci. 59:686-694.
Tinkle, D. W.   1969.   The concept of reproductive effort and its
     relation to the evolution of life histories of lizards.   Amer.
     Natur. 103:501-516.
Tinkle, D. W., H. M. Wilber, and S. G. Tilley.   1970.   Evolutionary
     strategies in lizard reproduction.   Evolution 24:55-74.

THE ORGANISATION OF CHALCID-DOMINATED PARASITOID COMMUNITIES

CENTRED UPON ENDOPHYTIC HOSTS

R.R. Askew

Department of Zoology, University of Manchester

England

The phytophagous larvae of many small endopterygote insect species complete their development inside plant tissue. Some inhabit stems or floral parts, others mine leaves and the most specialised induce hypertrophy of plant tissue to produce galls. Such endophytic species enjoy an environment largely protected from climatic extremes and one that is seldom breached by predators. Endophytes, however, are usually very vulnerable to the attack of parasitoids and they can support extremely complex parasitoid communities. These communities are self-contained in the sense that scarcely any parasitoids (or parasites as they will henceforth be termed) attack more than one group of endophytic hosts. The oak gall community is qualitatively almost totally different from the deciduous tree leaf-miner community, although both communities are part of the woodland ecosystem.

Interrelationships within an endophytic community are complex but, because activities occur within a very restricted space in which each occupant leaves some trace of its presence, it is usually not difficult to ascertain the specific rôle of each member of the community. Food webs are broad rather than long; many species may be included but since these are all more or less closely allied taxonomically they do not represent a variety of trophic levels.

It is proposed to describe two examples from Britain of endophytic communities: those in cynipid oak (*Quercus*) galls and in certain deciduous tree leaf-mines. Both, like the majority of such communities, are dominated by chalcidoid parasites. By comparing these two communities, and examining similarities and differences in their organisation, it is possible to identify some of the strategies employed by the component chalcid species.

130

Parasite strategies in communities centred upon a gall midge and a sawfly have been examined, respectively, by Force (1972) and Price (1973a) who conclude that development of the parasite complexes is related to plant succession with the early colonizers being highly fecund and of low competitive ability (r-strategists) and the late colonizers less fecund but of high competitive ability (K-strategists). This succession was found to apply also within a single host generation, and it is this aspect that is examined here. The present work pertains to mature woodland and the communities investigated have a much higher species diversity than those studied previously.

THE OAK GALL COMMUNITY

At least 31 species of gall-making Cynipidae are associated with oak in Britain. Most of these have two generations during a year and the galls induced by the two generations of the same species differ considerably in structure and position. The fauna known to be associated with these galls includes 15 species of inquiline cynipid and 45 species of chalcid parasite, 41 of the chalcid species being exclusively restricted to oak galls. Examination of large numbers of galls showed the community in any one type of gall to include at least some, and usually most, of the following elements (Askew 1961b).

1. gall-making cynipid feeding on gall tissue.

2. inquiline cynipid feeding on gall tissue and usually also destroying the gall-maker.

3. parasites of the gall-maker that are gall-specific. ⎫
4. parasites of the gall-maker that are not gall-specific. ⎬ 16 species

5. polyphagous parasite especially of inquilines and gall-makers but sometimes of other parasites, also consuming some gall tissue (only *Eurytoma brunniventris* Ratzeburg in this category).

6. gall-specific polyphagous parasites. ⎫
7. polyphagous parasites associated with a variety ⎬ 28 species
    of galls.

No instance of host-specificity involving a host other than a gall-maker was recognised. Examples of all of the above numbered categories are found in the food web existing in galls of *Andricus curvator* Hartig (Fig. 1). This is a relatively complex food web. In other galls the pattern may be simpler, but always the same

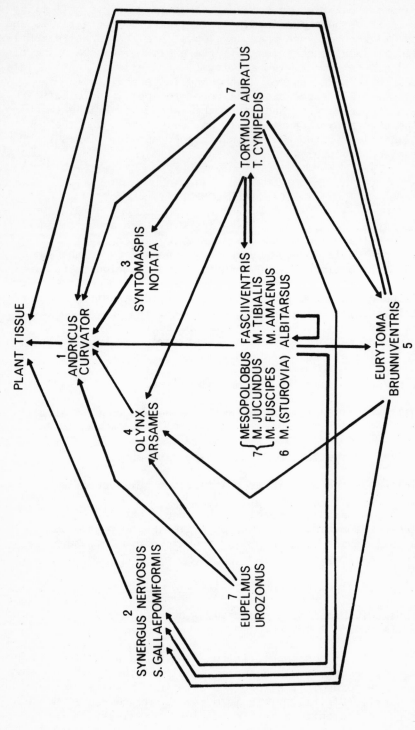

Fig. 1. Food web in galls of *Andricus curvator* Hartig. Arrows point towards the food source. Numbers refer to categories of gall inhabitants (see text).

species or their congeners occupy analogous positions in the community structure wherever they occur. Species assigned to different genera however, although taxonomically related, may occupy quite different situations (e.g. *Torymus* and *Syntomaspis*). This consistency of feeding habits within genera permits the construction of a generalised food web (Fig. 2), using data from all types of oak galls studied.

The five categories of parasites fall into two major groups; one composed of category 3 and 4 species which attack only the gall-makers and, for convenience, are all referred to as monophagous, even though those in category 4 are oligophagous, and the second of category 5, 6 and 7 species which are polyphagous. *Caenacis*, *Megastigmus* and *Tetrastichus* (also *Ormyrus* which is not included in Fig. 2) probably belong to this second group, but more data on their hosts are required.

What biological features characterise the two major groups of genera? Considering first the monophagous parasites, four families of Chalcidoidea are here represented but biological tendencies common to all, and distinguishing them from the polyphagous parasites, can be found. About half of the species are gall-specific (category 3) and the remainder (category 4) have very restricted host gall ranges, usually with a strong preference for only one or two types of gall. A striking feature of all of these species is that they make a very early attack on the gall-maker, usually achieving an initial high level of parasitism. Their flight periods are rather brief, very well synchronised with the appearance of their host galls, and they have only a single generation during a year, even though their host species may be bivoltine. Some of these features are illustrated by *Syntomaspis cyanea* (Boheman), a gall-maker parasite in galls of *Cynips divisa* Hartig (Fig. 3).

It is of interest to examine the course of parasitism in galls of the agamic generation of *Cynips divisa* through a year (Fig. 3). The galls, which are woody, globular, unilocular structures on the undersurfaces of oak leaves, first appear about the third week of June. They are very soon attacked by parasites and one of the first of these is *Syntomaspis cyanea*, the only species that restricts its attacks to the gall-making larva. *S. cyanea* is not strictly gall-specific, although its other recorded host galls all belong to the genus *Cynips*. The other parasite making an early attack on *C. divisa* galls is *Eurytoma brunniventris* which is a polyphagous species. The biologies of *S. cyanea* and *E. brunniventris* make an interesting comparison. *S. cyanea* is univoltine whilst *E. brunniventris* is multivoltine and it is mostly the second adult generation of the latter species that attacks young *C. divisa* galls. *S. cyanea* does not paralyse its host and feeds first on gall tissue, thus allowing the gall to

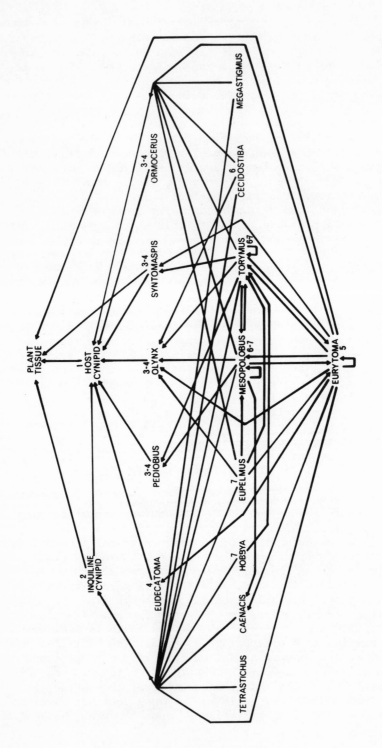

Fig. 2. Generalised diagram illustrating the interrelationships between genera of cynipid oak gall inhabitants. Arrows point towards the food source. Numbers refer to categories of gall inhabitants (see text).

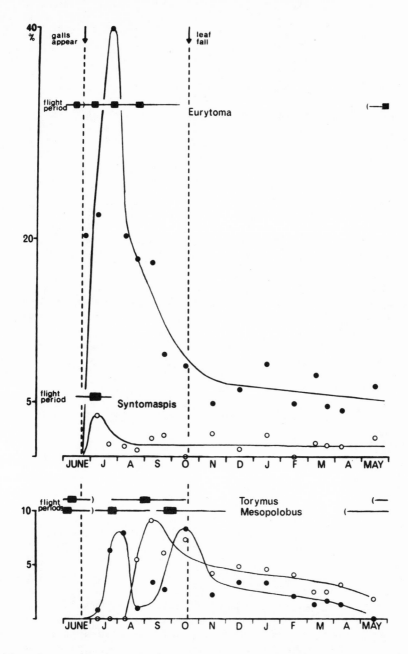

Fig. 3.   Seasonal changes in percentage occurrence of the major
parasites in galls of *Cynips divisa* Hartig based upon dissection of
3,281 galls collected in Wytham Wood, Berkshire, during 1958.
Curves have been fitted by eye.   Flight periods of the parasites
are indicated by horizontal lines, the generation peaks being shown
by solid rectangles.

develop fully. *E. brunniventris* usually paralyses the young *Cynips* larva which is soon eaten by the *Eurytoma* larva. This prevents further gall growth (Table 1). Since there is often insufficient food for the *E. brunniventris* larva in a young gall-maker larva, the *Eurytoma* frequently supplements its diet by also feeding on gall tissue, but this it does mostly after first consuming the gall-maker larva. *E. brunniventris* is the only polyphagous parasite known to eat gall tissue.

By allowing its host gall to develop, *S. cyanea* gains protection from attack by later parasites. It will be seen (Fig. 3) that *E. brunniventris* achieves a much greater initial parasitism of *C. divisa* than does *S. cyanea*, but that the numbers of *E. brunniventris* rapidly decline, partly as a result of the emergence of adult insects, but mostly because they are destroyed by later parasites such as *Mesopolobus* and *Torymus*. *S. cyanea* suffers to a much smaller extent from this mortality.

Olynx is another genus of monophagous parasites, the species of which, like *Syntomaspis*, do not paralyse their hosts but allow them to continue to grow and stimulate gall formation. Moreover, some species of *Olynx*, in some way unknown, alter the gall's structure so that it becomes harder and thicker-walled than normal (Askew 1961a).

In addition to benefiting from the protection of a fully-formed gall, by allowing the gall wasp larva to grow *Olynx* and *Syntomaspis* are provided with a much greater supply of animal food than would be the case if their host was paralysed by the ovipositing chalcid. In delaying destruction of their hosts, they are unusual ectoparasites. Typically, ectoparasites sting and either kill or paralyse the host at the time of oviposition. They thereby prevent the host from shedding its integument, which could be an awkward hazard to an ectoparasite, or from dislodging the ectoparasite by body movements. *Olynx* reduces the risk of dislodgement by placing its egg in the concavity of the ventral surface of its host (Askew 1961a).

By completing their development on the host stage originally attacked, typical ectoparasites do not enjoy the benefit of additional food becoming continually available to them in a host continuing to feed. Endoparasites, in contrast, seldom paralyse their host but allow it to continue feeding and growing whilst they initially consume only replaceable tissue. Perhaps the main difficulty in endoparasitism is that it necessitates overcoming the host's internal physiological defence mechanism. Endoparasitism is very rare amongst oak gall parasites but is found in *Eudecatoma* and *Pediobius*. Both of these genera are monophagous parasites and they make use of endophagy in perpetrating an early attack on their host populations.

Table 1. Sizes of *Cynips divisa* galls containing respectively, the gall-maker, *Syntomaspis cyanea* and *Eurytoma brunniventris*, showing that *S. cyanea* allows the gall to develop fully whereas parasitism by *E. brunniventris* results in a stunted gall. Data collected in Wytham Wood, Berkshire, 1958.

Contents of gall	Number measured	Mean diameter (mm)	Wall thickness of gall of mean diameter (mm)
Cynips divisa	7	4.7±0.15	0.9
Syntomaspis	14	4.7±0.13	1.0
Eurytoma	10	3.0±1.03	0.3

Another feature of monophagous parasites is that females tend to carry a large number of mature eggs (Table 2), an indication of high reproductive potential that in two species appears to be augmented by thelytoky, a phenomenon not associated with any of the polyphagous parasites. Males of *Olynx euedoreschus* (Walker) are exceedingly scarce; those of *Ormocerus vernalis* Walker are unknown.

Characteristics of polyphagous parasites in the oak gall community are very often complementary to those of species restricted to attacking the gall-making larvae. Most are multivoltine (only two of the *Mesopolobus* species are univoltine), and there is a positive correlation between the number of types of gall attacked by a species and the number of generations it passes through during a year (Table 3). As well as being polyphagous within a particular gall, these species generally each attack a wide range of galls. All are ectoparasites and, with the exception of *Eupelmus urozonus* Dalman, they nearly always paralyse their hosts at the time of oviposition so that, when attacking the gall-maker, gall development is curtailed. Since their flight periods are not synchronised with particular galls, these tend to be more prolonged than are those of the monophagous parasites (although an individual's life span is not necessarily any longer), and in species with three or more generations during a year, flight periods of consecutive generations may overlap.

Females of polyphagous parasites usually carry fewer mature eggs than do those of monophagous parasites (Table 2). None of

Table 2.   Egg loads of gravid chalcid parasites in cynipid oak galls
and *Phyllonorycter* leaf mines. Only fully-developed or
almost fully-developed eggs have been counted. All speci-
mens were collected in the field at Abbots Moss, Chesire,
in 1974.

PARASITES IN OAK GALLS	Number gravid	Mean egg no./ gravid female	Highest egg nos.	*Mean egg length
Gall-maker parasites				
Olynx skianeuros	6	81.5	134,112	0.22±.01
O. arsames	2	38.0	42,34	0.30
O. gallarum	2	20.0	24,16	0.15
**Pediobius lysis*	7	40.6	60,58	0.09±.03
Syntomaspis apicalis	6	21.7	30,28	0.50±.03
Polyphagous parasites				
Mesopolobus tibialis	27	12.1	30,29	0.41±.04
M. fuscipes	3	9.3	14,10	0.40
M. fasciiventris	14	9.7	18,18	0.43±.02
M. jucundus	4	15.3	28,12	0.50±.06
Cecidostiba semifascia	8	7.0	13,11	0.50±.04
Eurytoma brunniventris	9	10.9	20,17	0.30±.03
Megastigmus dorsalis	4	4.0	6,5	0.34±.04
Torymus auratus	22	15.6	30,24	0.47±.04
T. cynipedis	4	7.5	15,8	0.59±.07
PARASITES OF *PHYLLONORYCTER*				
Endoparasites				
***Enaysma niveipes*	14	20.4	34,32	0.15±.02
***E. latreillei*	12	29.1	45,42	0.15±.03
***E. cilla*	7	33.3	54,44	0.14±.02
***E. atys*	3	24.0	32,24	0.13
Pediobius alcaeus	10	10.6	18,17	0.28±.03
Chrysocharis nephereus	14	8.1	28,14	0.27±.03
C. phryne	3	7.7	12,7	0.22
C. laomedon	4	8.5	17,8	0.33±.03
Closterocerus trifasciatus	7	5.9	12,10	0.23±.03
Ectoparasites				
Cirrospilus diallus	17	6.6	14,10	0.40±.03
C. lyncus	5	6.8	15,8	0.34±.01
C. vittatus	3	5.0	10,4	0.24
Pnigalio longulus	23	7.8	23,21	0.46±.05
P. soemius	8	7.4	13,11	0.39±.04
P. pectinicornis	14	7.6	17,12	0.44±.05
Sympiesis gordius	2	4.5	5,4	0.40
S. sericeicornis	15	8.2	22,12	0.45±.05

*Excluding pedicel if present. Length in mm ± standard deviation.
**The very small eggs of these species were difficult to count and
the numbers given are liable to greater inaccuracy than is the case
for other species.

Table 3.   The relationship between the number of generations per
            year and the number of types of oak gall attacked in
            those chalcid species whose voltinism is certainly known.

Number of generations per year	Number of chalcid species	Mean number of galls attacked per species
1	11	3.2
2	10	5.7
2-3	2	9.5
3-4	2	13.5
4-5	1	19.0

the polyphagous parasites is thelytokous, and in several the sex
ratio of adults emerging from galls is male-biased. Considering
all of the oak gall parasites for which adequate data are available,
there appears to be a tendency for the sex ratio to become more
male-biased the more types of galls are attacked (Fig. 4). Although
the relationship is not close, there are significantly ($P<0.05$)
more species with less than sixty percent females amongst those
attacking more than five types of galls than amongst those
attacking fewer than five types of galls.

Discussion

     In the description of oak gall parasites, I have attempted to
show that the chalcid species adopt one of two broad strategies.
In the first of these the host range is restricted to the gall-
making cynipid and this is frequently accompanied by gall-
specificity. This trophic restriction necessitates early attack
on the gall-maker population before it is decimated by polyphagous
parasites and, accordingly, we find tightly host-synchronised
life cycles and adaptations to avoid possible food shortage as a
result of attacking very small hosts. Because the host galls are
small and soft when attacked, penetration is relatively easy and
oviposition rapid. These parasites carry large numbers of eggs
but often suffer considerable mortality from attack by polyphagous
parasites. In some ways they may be thought of as *r*-strategists
although, unlike the typical *r*-strategist parasitoid (Price 1973a),

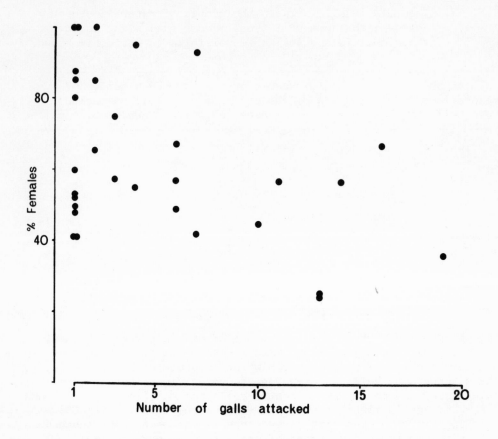

Fig. 4. Relationship between the sex ratios of emerging adults and
the number of types of gall attacked for species of chalcid parasites in cynipid oak galls. Only species for which at least fifteen
rearing records are available are included, and the majority of
ratios are based upon much larger numbers.

they are not very dispersive, seldom being found away from the
vicinity of hosts.

A quite different strategy is employed by the polyphagous
parasites in oak galls. These attack hosts in galls that are
usually fully-formed and therefore spend more time and energy
in the deposition of each ovum than do monophagous parasites.
Eurytoma brunniventris takes about one minute to penetrate a
young gall of *Cynips divisa* but this time increases up to about
20 minutes in the case of mature galls (Askew 1961b). An ability
to develop successfully on any host encountered is clearly

important. Polyphagous parasites carry fewer eggs on average than
do the monophagous parasites, but this indication of lower reproductive potential is counterbalanced by their multivoltinism and
by reduced mortality from hyperparasites as a result of their later
attack on host galls. These polyphagous parasites have some of the
attributes of K-strategists. It should however be noted that,
because of the succession of potential hosts within a gall, a lower
reproductive potential is less likely to be an adaptation to reduced
host density than a response to the increased energy cost of laying
each egg and the higher larval survival rate. The tendency for the
sex ratio of polyphagous parasites to be less female-biased than
that of monophagous parasites must also lower the reproductive
potential.

    A typical parasitoid K-strategist is characteristically
prevalent in a mature ecosystem; it is a climax species. The
polyphagous parasites depart from this ideal in apparently being
less restricted to the environment in which their hosts are found
than are the specific parasites. A collection of over six thousand
chalcids from sand-dunes some considerable distance from the nearest
oak tree included both sexes of five species of oak gall parasites
(*Torymus auratus* (Fourcroy) and four species of *Mesopolobus*) all
of which are polyphagous. Females of these species, carrying a
relatively light burden of eggs, would seem more suited to
dispersion than the heavily gravid females of the more host-
specific gall-maker parasites. The polyphagous parasites in oak
galls further depart from the ideal of a K-strategist in being
multivoltine; paradoxically the more host-specific parasites,
which are considered to be placed towards the r end of the r-K
continuum, are univoltine.

                    THE LEAF-MINE COMMUNITY

    Leaf-mines on deciduous trees may be formed by the larvae of
certain Diptera, Coleoptera, Hymenoptera and Lepidoptera, but the
chalcid parasites of these very different hosts, particularly
those of the last three orders, are similar. Although many leaf-
mining species may be identified by details in the form of their
mines, there is no diversity of mine form comparable to the great
variety of galls. The variation between mines involves little
more than differences in mine outline and the track followed by
the miner. All mines are essentially cavities, bordered on one or
both sides by an intact layer of epidermis. Different species,
however, mine different trees, and this provides a dimension to
niche diversity in some ways comparable to the variety of forms
exhibited by cynipid galls on oak. Qualitatively and quantitatively,
the most similar leaf-miner parasite faunas are found on taxonomically
allied trees (Askew and Shaw 1975); the most similar faunas in oak

galls are associated with those galls that are most alike in
structure, position on the tree and season of growth (Askew 1961b).

The group of deciduous tree leaf-miners about which we have
most information (Askew and Shaw 1975) is the genus *Phyllonorycter*
(= *Lithocolletis*). All species of this lepidopteran genus con-
struct very similar mines. Each of the numerous species is usually
confined to a single tree genus or to allied tree genera, although
several species may be found on the same tree.

The community of insects associated with *Phyllonorycter* mines
on trees is, in many ways, strikingly different in structure from
the oak gall community. Forty-six chalcid species were reared as
parasites of *Phyllonorycter*, together with three species of
Braconidae. These latter are present only at low densities and
will not be considered further here. The interrelationships between
chalcid genera, so far as are presently known, are indicated in
Fig. 5. These are complex and, at first sight, chaotic, compared
to the more structured, although equally large, oak gall community.

Endoparasites and ectoparasites are present in roughly equal
numbers (21:25) in leaf-mines. The ectoparasites attack primary
host larvae irrespective of whether or not they already contain
endoparasites, and they thus inflict mortality on the endoparasites
as well as on the leaf-miner population. Sometimes they destroy
other ectoparasites, either by parasitising them directly or, more
commonly, as a result of multiparasitism. Endoparasites attack
predominantly the leaf-miners, but occasionally also ectoparasites,
or other endoparasites usually at least after these have left their
hosts' bodies. We have also recorded a few instances of endoparasites not being stung by an ovipositing ectoparasite and
emerging fully fed from the primary host to leave the ectoparasite
to starve. Even though there is this variety of possible interactions, hyperparasitism by endoparasites is less frequent than
it is by ectoparasites. Only *Elachertus* among the ectoparasites
has not so far been found as a hyperparasite, whereas of the
endoparasites the several *Enaysma* species are apparently never
hyperparasitic and, with the exception of *Closterocerus*, the
others are only infrequently so (Table 4). This is just what would
be expected. Endoparasites tend to be more restricted than
ectoparasites in the hosts in which they can successfully develop
because the means by which they overcome the host's defensive
encapsulatory reaction do not usually operate over a wide host
spectrum. *Closterocerus* is exceptional in being an endoparasite
which evidently specialises to some extent as a hyperparasite
(Table 4). Work at present in hand may reveal other specialisations.

Host-specificity is unusual amongst leaf-miner parasites and
is developed to a high degree only in the endoparasitic genus

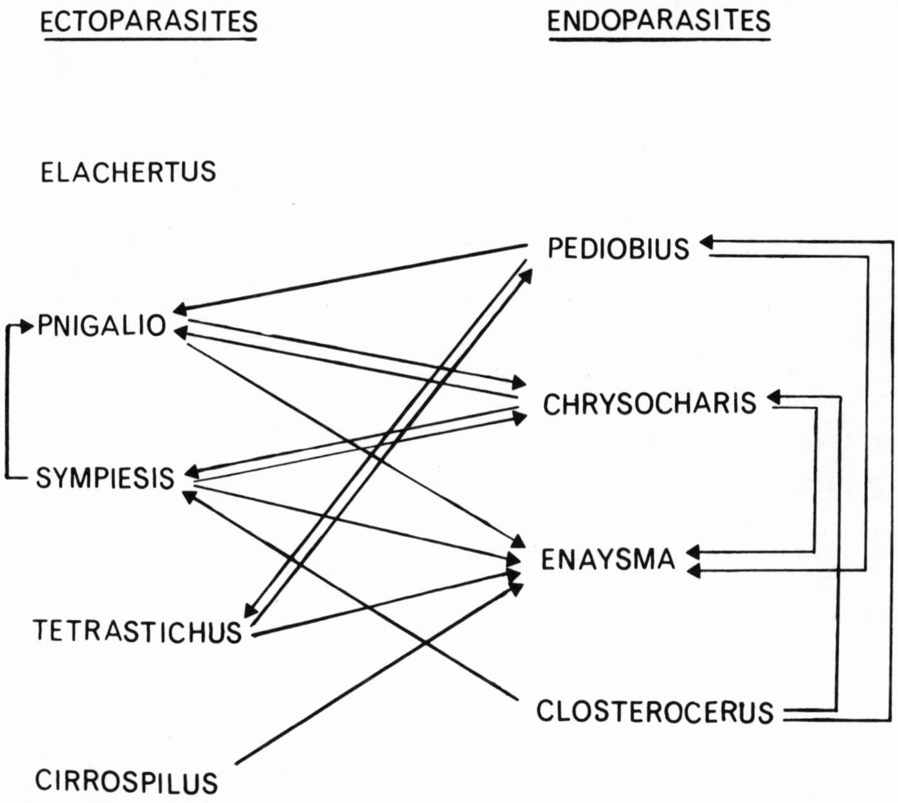

ECTOPARASITES   ENDOPARASITES

ELACHERTUS

Fig. 5.   Interrelationships between chalcid genera parasitic upon
species of *Phyllonorycter*.   Arrows point towards the food source.
All genera also attack *Phyllonorycter*.   Data collected in
Cheshire, 1973 and from Delucchi (1958).

Table 4.   Detected incidences of hyperparasitism by some
           Phyllonorycter parasites, based upon host remains in
           leaf-mines producing adult chalcids.   Cases of
           hyperparasitism by ectoparasites, in which the
           lepidopterous primary host contains young endo-
           parasites, would not be detected in this analysis,
           and the real percentage hyperparasitism by ecto-
           parasites is certainly greater than that indicated
           by these data.   The figures pertain to oak and birch
           mines collected in Cheshire, 1973.

	Number developing as:		
	a)primary parasites	b)hyper- parasites	% hyper- parasitism
ECTOPARASITES			
Cirrospilus diallus	19	3	13.6
Sympiesis sericeicornis	40	4	9.1
S. gordius	13	0	0
Pnigalio longulus	38	2	5.0
P. pectinicornis	48	6	11.1
Tetrastichus ecus	11	6	35.3
ENDOPARASITES			
Pediobius alcaeus	87	1	1.1
Chrysocharis laomedon	65	6	8.5
C. nephereus	34	6	15.0
Enaysma latreillei	21	0	0
E. niveipes	24	0	0
Closterocerus trifasciatus	12	17	58.6

Enaysma. Species of *Enaysma* probably attack only *Phyllonorycter* larvae and each species is virtually restricted to hosts mining only one or a very limited number of tree genera (Askew and Ruse 1974a). Some other endoparasites have restricted primary host ranges. Two species of *Pediobius*, *Chrysocharis gemma* (Walker) and *C. phryne* (Walker) are all associated with narrow ranges of trees; there is evidence, in fact, that *Pediobius alcaeus* (Walker) consists of two host tree-defined biological forms, one of which is thelytokous. Likewise, two tree-defined forms of *Chrysocharis nephereus* (Walker) have been recognised (Askew and Coshan 1973). Amongst the ecto-parasites, a comparable situation exists only in the two sibling species *Sympiesis sericeicornis* (Nees) and *S. grahami* Erdös which, again, are predominantly associated with different trees.

As the endoparasites of *Phyllonorycter* tend to be more restricted in their host ranges than the ectoparasites, we might expect them to have some of the same biological characteristics as the specific cynipid parasites in oak galls.

Univoltinism, a feature of specific cynipid parasites, is unknown amongst *Phyllonorycter* parasites, possibly because mine form remains constant in different host generations.

The endoparasites of *Phyllonorycter*, like those of oak Cynipidae, tend to attack earlier host stages than the ectoparasites (Table 5). However, because leaf-miner development is rapid and mines at vary-ing stages of development are present together over a long period, this does not demand that the endoparasites have an earlier flight period than the ectoparasites.

Two other features in which the endoparasites in leaf-mines show parallel adaptations to specific gall wasp parasites are egg load and sex ratio. The largest egg loads are carried by the endopara-sitic *Enaysma* species (Table 2), and this can be directly related to their extremely narrow host ranges. If this is an indication that *Enaysma* are channeling their energies towards a high reproductive output, it would substantiate the proposition that they are *r*-strategists attacking an early, abundant host stage. Price (1973b) reports that ectoparasitic Ichneumonidae have fewer ovarioles per ovary than endoparasitic species.

The more host tree-restricted parasites of *Phyllonorycter* tend to have a more female-biased sex ratio (Fig. 6), and this must further contribute to their overall fecundity. *Chrysocharis phryne*, in fact, is probably thelytokous. The correlation between female bias and host range size is rather better than in the case of para-sites in oak galls (Fig. 4), but this, to a large extent, is because species of the single genus *Enaysma* have a consistently small number of host trees and a high percentage of females.

Table 5.   The larval instars of *Phyllonorycter* mining *Betula*
           attacked by endoparasites and ectoparasites, Abbots
           Moss, Cheshire 1974.   Data obtained by examining mine
           contents; a *Phyllonorycter* larva containing living
           endoparasite eggs or first instar larvae was scored
           as having been originally attacked in the same instar,
           but larvae containing more advanced endoparasites were
           ignored; the figures for endoparasites thus represent
           the latest possible stage attacked.   The presence of
           ectoparasites of any age was used to provide data on
           the instars attacked by ectoparasites.

| Number of larvae attacked by: | *Phyllonorycter* larval instar attacked | | | | | Totals |
	1	2	3	4	5	
Endoparasites	12	58	29	6	1	106
Ectoparasites	0	6	28	66	73	173

Discussion

Compared with galls, leaf-mines are flimsy structures.   They
are easily damaged and persist intact upon the trees for a
comparatively short time.   This probably accounts in part for the
less ordered temporal sequence of parasite species in leaf-mines
than in oak galls.

Leaf-miners do not modify plant tissue in the manner of gall-
makers; they merely destroy it.   There is no provision in leaf-
mines for phytophagy by the parasites and we find no early-attacking
ectoparasites in leaf-mines, supplementing their diet with plant
food, like *Syntomaspis* and *Eurytoma* in oak galls.   As the ectoparasites sting their hosts, they cannot rely upon host growth
to supply food for their offspring, and inevitably the younger
larval stages of *Phyllonorycter* are much more heavily attacked
by endoparasites than by ectoparasites.   Conversely, ectoparasites
concentrate their attack upon older hosts.   Since both endoparasites
and ectoparasites are ovipositing at much the same time, this
divergence in host selection must reduce, to a certain extent,
the likelihood of an ectoparasite attacking a host already

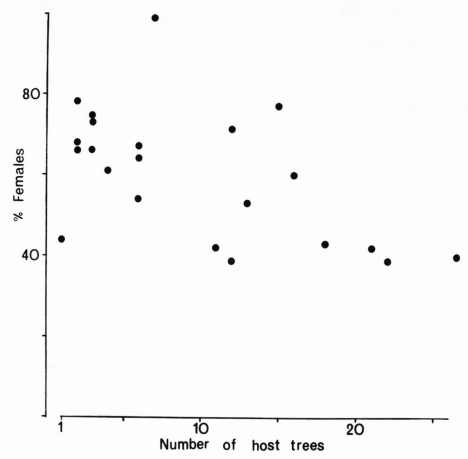

Fig. 6. Relationship between the sex ratios of emerging adults and the number of trees with which associated for species of chalcid parasites of *Phyllonorycter*. Only species for which at least fifty rearing records are available are included.

containing an endoparasite. If development of individuals in a *Phyllonorycter* population was more synchronised and the endoparasites were compelled in consequence to have a significantly earlier flight period than the ectoparasites, a greater proportion of them would presumably be later destroyed by ectoparasites, especially since the structure of a leaf-mine, unlike that of a gall, does not confer increasing protection upon its inhabitants as it matures. If the endoparasites were able to attack late larval stages of *Phyllonorycter*, they would again reduce their losses from ecto-parasitic attack, but late larval stages of the host are at a relatively low density and we have no certain evidence that this policy has been adopted by any species (although *Pediobius alcaeus* may have done so). A further factor mitigating against this strategy is that late-attacking endoparasites would be at risk from endo-parasites that had previously attacked the host population since the oldest endoparasite usually survives in cases of multiendoparasitism (the converse is true in multiectoparasitism). For an endoparasite, the most likely alternative tactic to that of attacking the host at an early state of its development would seem to be that employed by *Closterocerus*. *Closterocerus* has escaped the bonds of host-specificity (theoretically, at least, difficult for an endoparasite) and become a late-attacking, polyphagous parasite.

Since host scarcity is not likely to be a serious problem for polyphagous parasites in oak galls, their smaller average egg load, compared with that of the specific parasites, is attributed to the high energy expenditure in ovipositing in mature galls (see above). In leaf-mines, however, the situation is different. Mines may be opened by insectivorous birds and there is much destruction of *Phyllonorycter* larvae by host-feeding chalcids. Thus more and more mines are without living larvae as the season progresses and, accordingly, the late-attacking polyphagous parasites carry few eggs, even though the energy cost of attacking a mature mine (excluding the cost of searching for an inhabited one) is probably no greater than that of attacking a young mine. In other words, polyphagous parasites in galls and mines are limited in the number of eggs they can successfully place in a given time but for different reasons. In galls the limitation is imposed by gall structure, in mines by host scarcity.

GENERAL DISCUSSION

Both endophytic communities described in this paper are components of mature ecosystems. Price (1973a) suggests that species in such communities may not clearly show the *r*-strategem, and this view is largely supported by the data presented here. It should be remembered, however, that nearly all of the parasites belong to a limited taxonomic group, and this no doubt places a restraint upon extreme divergence of many of their biological characteristics. It is encouraging that certain strategies are discernible and these, in most although not all features, conform in a general way to the concept of *r*- and *K*-strategies.

Parasites attacking an endophytic host in the early stages of its development find a comparatively high host density but the host individuals are small. Among such parasites we find those carrying a large number of eggs (e.g. *Olynx*, *Enaysma*), and they are either endoparasites allowing their hosts to continue their development or, in the case of some of the oak gall parasites, they are ectoparasites also allowing continued host growth and, in a few species, supplementing their diet with gall tissue.

A major hazard confronting species attacking an endophytic host early is that of later destruction by secondary parasites. The communities become more and more complex as the season progresses and more species are included. Because the original plant-feeding host is soon decimated, and an ever-increasing proportion of live larvae in the plant cells are parasites, later parasites tend to be polyphagous, able to develop on either the primary host or its parasites. The polyphagous habit is of particular importance to oak gall parasites because time and energy spent penetrating galls in search of a selected host species would probably be considerable. We have no evidence that chalcid parasites in either leaf- miners or cynipid oak galls avoid drilling with their ovipositors into cells containing previously parasitised hosts. This apparent inability to determine the contents of a cell from the outside is perhaps one reason why no parasite in either community certainly specialises as a hyperparasite of a particular primary parasite species or even as a late-attacking primary parasite of the phytophagous host only. The only case of specialisation by parasites for a host other than the original leaf-miner or gall-maker that I am aware of is parasitisation of the inquiline cynipid *Periclistus* by *Eurytoma rosae* Nees and *Caenacis inflexa* (Ratzeburg) in rose galls of *Diplolepis rosae* (L.) (Claridge and Askew 1960, Callan 1944).

Polyphagy is generally associated with ectoparasitism, and accordingly we find that the parasites attacking older host stages usually feed in this manner. Multiparasitism must reinforce the tendency for endoparasites to attack the host early and for ectoparasites to attack it later. When larvae of two endoparasite species occur in the same host individual it is usually the older one that survives. In contrast, a younger ectoparasite normally develops at the expense of an older ectoparasite on the same primary host. Another factor that may encourage endoparasites to attack young host stages is that encapsulatory defensive reactions are probably more potent in old than in young host larvae.

The fundamental distinction between an *r*- and a *K*-strategist is reflected in their respective intrinsic rates of natural increase. These have not been ascertained for the parasites discussed in this paper, but I have used the mature egg load of gravid females caught in the field as a probable indication of reproductive potential. Those parasites attacking oak galls early often carry a large number of mature eggs. This is not so marked amongst leaf-miner parasites, only the very host-restricted

Enaysma species carrying large numbers of eggs.

An attempt has been made to present, in simplified form, the
several factors that seem to be of significance for parasites
attacking young and old host stages, and the strategies most
frequently adopted by the parasites in response (Fig. 7). With
these in mind, it is possible to speculate on the evolution of
the type of community described.

It is difficult to conceive of a specific endoparasite of an
endophytic host establishing itself in a community already rich in
ectoparasites. Precise and perhaps complex adaptations would be
required. It is much easier to envisage endoparasites becoming
adapted to attacking an endophytic host before the appearance of
ectoparasites in the community. However, the protected endophytic
situation would, sooner or later, attract ectoparasites and competition between endoparasites and ectoparasites would ensue.
Selection would then favour the adoption of more defined *r*-
strategies by the endoparasites and *K*-strategies by the ectoparasites for reasons already discussed. If this generalisation
is correct, an indication of the maturity of an endophytic
community may perhaps be obtained by the proportion of ectoparasites in its fauna. The oak gall community includes a very high
percentage of ectoparasites, far in excess of the percentage
in the leaf-miner fauna, which would suggest that it is the more
mature community of the two. A feature of the oak gall community
consistent with this view is its taxonomic diversity, chalcids
belonging to six families being represented. In contrast, the
leaf-miner community is dominated by one family, the Eulophidae.
The relatively few Eulophidae established in the oak gall community
(*Olynx, Pediobius*) are taxonomically isolated there and have well-
developed *r*-characteristics. It is tempting to think of them as
the relic of a once more extensive eulophid element that has been
decimated by the arrival of more competitive, ectoparasitic *K*-
strategists. Force (1972) has postulated an increasing dominance
of *K*-strategists in parasitoid communities as succession proceeds.

Arguing along these lines, a third endophytic community, that
centred upon cecidomyiid galls on birch (*Betula*) leaves (Askew and
Ruse 1974b), is perhaps at a relatively young evolutionary stage:
endoparasites dominate its parasite fauna which includes no fewer
than six species of the eulophid genus *Tetrastichus* amongst the
ten chalcid species recorded.

SUMMARY

The biological strategies adopted by the parasitic chalcid
components of two endophytic communities, those in cynipid oak

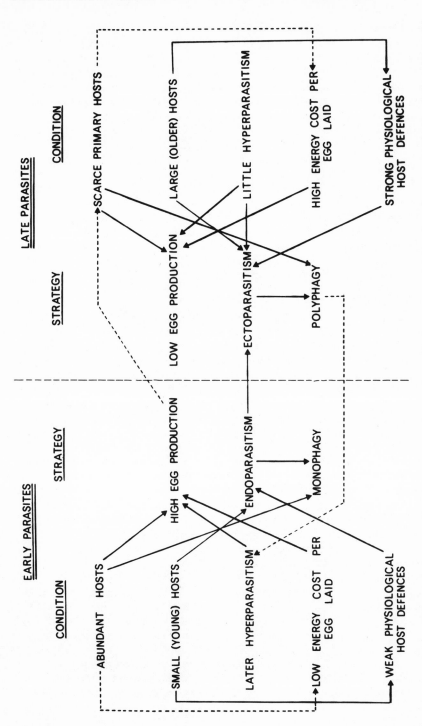

Fig. 7. Simplified scheme of basic strategies adopted in response to various conditions by chalcid parasites attacking respectively young and older host stages in endophytic communities. Most parasites in the two communities examined in this paper conform to this pattern; those that do not have special adaptations.

galls and in *Phyllonorycter* leaf-mines, are described and discussed.
In both communities species tending towards either end of the *r*-
K-strategy continuum can be identified. Those thought of as
r-strategists are characterised by being monophagous, often endoparasitic, and carrying a large egg load; those with *K*-strategist
tendencies are polyphagous, ectoparasitic, and carry a smaller egg
load. Biotic factors favouring the adoption of these two sets of
characteristics are outlined (Fig. 7), and the possible pattern of
evolution of parasite communities is discussed.

ACKNOWLEDGEMENTS

I am indebted to the Natural Environment Research Council
whose grant (GR 3/864) made the time-consuming studies on the
leaf-miner community possible, and to Dr. M.R. Shaw whose assist-
ance, advice and ideas were invaluable. I am grateful also to
Mr. J.G. Blower for criticism of a draft manuscript.

LITERATURE CITED

Askew, R. R. 1961a. The biology of the British species of the
    genus *Olynx* Förster (Hymenoptera: Eulophidae), with a note
    on seasonal colour forms in the Chalcidoidea. Proc. R.
    Entomol. Soc. Lond., Ser. A. 36:103-112.
Askew, R. R. 1961b. On the biology of the inhabitants of oak
    galls of Cynipidae (Hymenoptera) in Britain. Trans. Soc.
    Br. Entomol. 14:237-268.
Askew, R. R., and P. F. Coshan. 1973. A study of *Chrysocharis
    nephereus* (Walker) (Hymenoptera: Eulophidae) and allied
    species, with observations on their biology in Northern
    England. J. Nat. Hist. 7:47-63.
Askew, R. R., and J. M. Ruse. 1974a. Biology and taxonomy of
    species of the genus *Enaysma* Delucchi (Hym., Eulophidae,
    Entedontinae) with special reference to the British
    fauna. Trans. R. Entomol. Soc. Lond. 125:257-294.
Askew, R. R., and J. M. Ruse. 1974b. The biology of some Cecido-
    myiidae (Diptera) galling the leaves of birch (*Betula*) with
    special reference to their chalcidoid (Hymenoptera) parasites.
    Trans. R. Entomol. Soc. Lond. 126:129-167.
Askew, R. R., and M. R. Shaw. 1975. An account of the Chalcidoidea
    (Hymenoptera) parasitising leaf-mining insects of deciduous
    trees in Britain. J. Linn. Soc. In press.
Callan, E. McC. 1944. *Habrocytus bedeguaris* Thomson and
    H. periclisti sp.n. (Hym., Pteromalidae) reared from galls
    of *Rhodites rosae* (L.). Proc. R. Entomol. Soc. Lond., Ser. B.
    13:90-93.

Claridge, M. F., and R. R. Askew. 1960. Sibling species in the *Eurytoma rosae* group (Hym., Eurytomidae). Entomophaga 5: 141-153.

Delucchi, V. 1958. *Lithocolletis messaniella* Zeller (Lep. Gracilariidae): Analysis of some mortality factors with particular reference to its parasite complex. Entomophaga 3: 203-270.

Force, D. C. 1972. *r*- and *K*-strategists in endemic host-parasitoid communities. Bull. Entomol. Soc. Am. 18:135-137.

Price, P. W. 1973a. Parasitoid strategies and community organisation. Environ. Entomol. 2:623-626.

Price, P. W. 1973b. Reproductive strategies in parasitoid wasps. Amer. Natur. 107:684-693.

INTERACTIONS OF SEEDS AND THEIR INSECT PREDATORS/PARASITOIDS

IN A TROPICAL DECIDUOUS FOREST

Daniel H. Janzen

Department of Biology, University of Michigan

Ann Arbor, Michigan 48104, U.S.A.

In this paper I will touch on some of the patterns and processes that have come to light in an on-going study of how the
insects that eat seeds in tropical vegetation influence the
structure of that vegetation and the characteristics of its individual members, and how the characteristics of the plants influence
the insects. This study began in 1964 in the lowlands of the state
of Veracruz, Mexico. While trying to understand the population
dynamics and interdependency of obligate acacia-ants and their
acacias (Janzen 1967), I was struck by the thoroughness with which
the bruchid *Acanthoscelides oblongoguttatus* decimated the seed
crops of *Acacia cornigera*. This swollen-thorn acacia was extremely
common as a result of farming and grazing practices, and it produced
far more fruits than the local community of dispersal agents would
eat (a thorough study of dispersal agents was not made, but at
least the Plain-tailed Brown Jay, *Psilorhinus mexicanus*, Black-
headed Saltator, *Saltator atriceps*, and Grayish Saltator, *Saltator
coerulescens*, ate seeds along with the pulp and defecated living
seeds). While most of the *A. cornigera* population bore flowers
in the later part of the dry season (March, April) and mature fruits
about ten months later, individuals in almost any reproductive
stage could be found at any time of year in an area as small as a
few hectares. The pods remained on the acacia in a stage susceptible to bruchid oviposition for as long as two to three months.
Attack was heavy and continuous, both from beetles that could have
emerged from other pods on the same acacia, and from beetles coming
from other *A. cornigera* out of fruiting synchrony. Only very broad
synchrony of flowering and fruiting occurred in the highly mixed
and variously disturbed habitats (roadsides, back yards, various
age corn fields, pastures, fencerows, creekbanks, marshes).

154

It was obvious that in the highly disturbed lowlands of Veracruz, the acacia was largely losing the race between the bruchids and the dispersal agents. However, it seemed likely that in a more intact habitat where *A. cornigera* would be much rarer, more synchronized, and further apart, the interaction would not be so one-sided. It was also obvious that such habitats were not available to me, so the subject was shelved while I worked on the interaction between the ants and the acacias.

While working out the ecological distribution of the ants and their acacias throughout Central America during July and August of 1966 and 1967 (Janzen 1974a), I had the opportunity to make general collections of legume seeds in many habitats. Confronted with numerous bags of seed pods, from some of which came large numbers of bruchids while from others came none, it was obvious that there was a pattern. Large-seeded legumes in general lacked bruchids while almost all of the small-seeded ones had bruchids (Janzen 1969). It is perhaps appropriate to add here that the Janzen (1969) paper, which attempts to relate seed size, seed toxicity, and bruchid attack, contains two quite annoying errors. First, after developing the entire paper around the idea that species preyed on by Bruchidae have smaller seeds, more seeds, and more biomass of seeds per unit canopy volume, the first sentence of point one in the "Conclusions" section contains a ludicrous printing error, such that the sentence reads just the opposite (the word "not" belongs between the *first* "are" and "attacked", not the second, page 23, column 1, lines 36-40). Second, the primary biological exception to the tentative rule put forth in the paper was a very small-seeded species of *Indigofera*; further collecting in Mexico and Central America has shown that this legume is heavily attacked by a very small species of bruchid (unpublished field notes, Johnson 1973, and Center and Johnson 1974).

The consideration of large seeds as toxic was nearly as serendipitous as finding that they lacked bruchid attack. I was walking down a hall in the Botany Department at the University of Kansas holding a Costa Rican cycad fruit in my hand, and a person walking the other way commented that he had a whole refrigerator full of them and did I know that they have very toxic seeds? I said no, I did not know that, but that I had a seed upstairs that I was willing to bet would be very toxic and did he wish to analyse it. In a few hours E. Arthur Bell was back with the announcement that my *Mucuna andreana* seed had an extraordinarily high concentration of L-DOPA (L-dihydroxyphenylalanine), an uncommon and non-protein amino acid with well known toxic physiological effects on mammals. This led to a number of examinations of large and bruchid-free seeds for high concentrations of potentially toxic secondary compounds (e.g., Bell and Janzen 1971) and to an increasing frustration with our lack of knowledge of whether these compounds really are toxic to bruchids.

Serendipity struck again. Back in 1966 I had met Paul Feeny while he happened to be passing through the University of Kansas, and he told me about his work with tannins in oak leaves and how he had fed them to armyworms (*Prodenia eridania*) in the laboratory to illustrate tannin toxicity. I had then collaborated with him and Sherry Rehr to test an old hypothesis (Janzen 1966, 1967) that the swollen-thorn acacia leaves would be less toxic to insects than their non-ant-acacia relatives; they are less toxic (Rehr et al. 1973a) and Feeny was an obvious person to test the L-DOPA and other compounds that Bell had found in large legume seeds. L-DOPA and other seed compounds were then found to be toxic to armyworms, depending on concentration (Rehr et al. 1973b,c).

Looking at the bruchids, their interplay with seed dispersal agents, and the toxicity of the seeds to some animals but not all, suddenly caused me to realize that these and other seed predators could be not only influencing the genetic characteristics of individual plant species in the habitat, but also influencing their relative abundance. By asking myself exactly what is it that decides whether an adult of tree species X actually appears at point Y in the forest, I found myself "reinventing" all those things that I should have learned about probability theory when I was an undergraduate. I spent about six months figuring out how to graphically represent the interaction between the size and dispersal pattern of a seed crop, and a seed predator that is differentially effective depending on the number of seeds present. I then asked Richard Levins about it, and he pointed out that the "model" I had invented was the answer to one of the questions on the final examination in mathematical biology that he was teaching that semester. At any rate, the resulting theoretical analysis of the potential impact of seed predators on forest community structure (Janzen 1970) forms the underlying philosophical structure for most of what follows here. Since that time, I have tried primarily to obtain a data base that would allow me to test some of the hypotheses presented in 1970, and have gathered data so as to optimize the discovery of patterns clarifying the questions posed in the first sentence of this paper. I will deal with a series of interrelated but not all-encompassing patterns and apparent hypotheses that have come to light by focusing on one major vegetation type, tropical deciduous forest, in one area, the Pacific coastal lowlands of Costa Rica (primarily Guanacaste Province). This particular study began in 1965 while teaching for the Organization for Tropical Studies and would not have come about without the logistic and intellectual support of that organization, and the students, faculty, and researchers working with it.

DEFINITIONS

There are only two bothersome terminological snarls of concern here: *seed predators* and *secondary compounds*.

Ecologically and behaviorally, the adult bruchid (or weevil, cerambycid, etc.) is acting as a predator (Janzen 1971a). It moves through the habitat searching out and killing individual seeds. It kills them by laying an egg on or near them. It is satiated by running out of eggs in a given time interval. On the other hand, the bruchid larva functions as a different sort of predator, one commonly called a parasitoid in the biological control literature. The larva kills (usually) only one prey individual, does it slowly, does not leave it, etc. From the standpoint of the adult tree, the plant is losing its seed to a seed predator as surely as if it were losing them to a deer or squirrel. From the standpoint of the ecol- ogy of the insect, it is traditional to view the insect as a para- sitoid. Therefore in discussions of tree ecology, I will think of them as seed predators, but in asking why there are a given number of them found on a host, it will be most profitable to think of them as parasitoids.

Secondary compounds are chemicals found in seeds (and in other plant parts) that are not part of the metabolic processes common to all or most plants. They are usually functional as floral and fruit attractants, antibiotics, repellants, and/or toxicants, and usually occur at much higher concentrations than do other odd compounds operating within the physiology of the plant (hormones, germination inhibitors, etc.). In this paper, my comments about secondary compounds are restricted to those that generally have a negative physiological effect on an animal (e.g. alkaloids, uncommon amino acids, terpenes, phenols, saponins, lectins, etc.).

DENSE VERSUS DIFFUSE POPULATIONS

A core aspect of the potential impact of pre-dispersal seed predators on trees is that as the distance between seed crops in space and/or time increases, the percent seed mortality within an individual tree's crop should decline. This hypothesis is extra- ordinarily difficult to test in nature, owing to the difficulty of interpreting the data obtained when comparing percent seed mortality among crops of clumped conspecifics with crops of diffuse- ly distributed conspecifics. To illustrate this difficulty, I here present a partial analysis of the predation on seeds of clumped and widely spaced *Acacia farnesiana* ("guisache", Mimosaceae) by two species of host-specific bruchids, *Mimosestes sallaei* and *M. immunis*. Background natural history is given before the actual test.

The study site lies along the northeastern edge of the approxi- mately 25 km^2 forested seasonal swamp behind the hill behind the Organization for Tropical Studies (OTS) Palo Verde Field Station, which is in turn in the southwestern end of the COMELCO ranch, which is in turn to the southwest of Bagaces, Guanacaste Province,

Costa Rica (about 10°40' N. Lat., 10 m elevation). *A. farnesiana* is very widely distributed in Central and northern South America, and indigenous to the study site; there is no hint that either species of *Mimosestes* is recently introduced. It is impossible to know how much European man, his cattle, and his lumbering have disturbed the site, but it appears to never have been cleared for field or pasture. However, isolated trees have been cut, firing history is undoubtedly different from the conditions under which the bruchid-acacia interaction evolved, and the site has been foraged in by semi-feral cattle for at least 100 years and perhaps as long as 300 years. It is about as natural a habitat for *A. farnesiana* as can be found in Costa Rica. The swamp and the adjacent forest still contain individuals of what may have been dispersal agents for *A. farnesiana* (white-tailed deer, peccaries, agoutis, pacas, and smaller rodents) but their density and relative proportions undoubtedly do not represent those under which the bruchid-acacia interaction evolved. Cattle now occasionally eat variable amounts of *A. farnesiana* pods and likewise serve as dispersal agents, but there is no way to know to what degree they replace the native animals as dispersal agents.

The study site is at the edge of the swamp, lying to the north of where the swamp edge is cut by the dirt track from the COMELCO main ranch headquarters to the Palo Verde Field Station. In those portions of the swamp not severely damaged by roadwork, fire, and cattle, the vegetation consists of a dense stand of *Parkinsonia aculeata* ("palo verde") in the center, surrounded by concentric variable-width rings of *Pithecellobium dulce*, *Cocoloba caracasana*, *Acacia farnesiana*, and *Acacia collinsii*. Each of these species is found abundantly mixed into the vegetation outside of the ring of its greatest density. As one approaches the edge of the swamp, the species richness of woody plants begins to climb very rapidly and the adjacent hillside may have as many as 150 species of trees, woody shrubs, and woody vines.

The *dense stand* of *A. farnesiana* lies approximately 500 m into the swamp and contains roughly 10 to 30 adults per hectare, with only scattered other species of low trees or shrubs mixed among them. Each *A. farnesiana* is about 1.5 to 3 m tall, with a crown of 5 to 30 m circumference. It is 2 to 30 m between their crowns. The *diffuse stand* of *A. farnesiana* is at the edge of the swamp; the acacias occur at a density of about 1 per hectare, with 50 to 100 m between conspecifics. The intervening spaces are filled with about 30 species of woody shrubs and small trees and it is generally impossible to see one *A. farnesiana* when standing at another.

In an earlier terminology, it would have been customary to speak of the dense stand as being the center of the population and the diffuse stand as the edge of the population. However, since we have not the slightest idea of where the seeds come from to

produce any given new acacia, nor what is the longevity of acacias
at either site, I am reluctant to apply any terms that carry implications about the spatial dynamics of the population. For all we
know, three adults in the diffuse site may contribute all the acacia
adults to be recruited to the population during the next hundred
years.

The two species of bruchids occur as seed predators on no other
species of plants in this habitat. The adults can, however, be taken
with a sweep net from the foliage of the habitat at any time of year
and both have been caught drinking nectar in the flowers of *Parkinsonia aculeata*. There are at least 80 other species of bruchids that
can be reared from seeds growing within a 1 km radius of the site;
none of these feed on *Acacia farnesiana* seeds.

A. farnesiana bears scattered flowers from July through March,
but the major flowering period is in the dry season at most sites.
The flowers are obligatorily outcrossed (if my experimental crosses
made in the lowlands of Veracruz in 1964 are representative) and the
pods mature by the middle of the following dry season. At the time
of pod collection in the present study (14 March 1971), about 50%
of the trees had dropped all their pods and on the remainder, most
pods were mature. The indehiscent and heavy pods fall straight to
the ground, where they are occasionally eaten entirely by deer,
peccaries and cattle, chewed up by rodents, or carried off by rodents. Apparently these animals are interested in the slightly
sweet pod wall, which may also contain a substantial amount of
nutrients if we can reason from the analyses of pods of other mammal-
dispersed acacias (Gwynne 1969). Some of the pods fall down the
cracks in the drying mud surface and others are probably washed
away from the parent plant by the torrential downpours that hit
this habitat during the early rainy season. It is doubtful if any
of the seeds consumed by mammals are killed by digestive action,
but rodents may chew through their hard seed coats.

The bruchids are present in the habitat and will oviposit on a
full-sized pod at any time of year, as shown by fresh eggs on pods
in miscellaneous collections of full-sized green pods made in and
near the swamp in June, July, August, September, December, February,
and March. This makes the plant's behavior, of holding the pod size
quite small until only a few months before it will mature, very
understandable. If full-sized pods were present from flowering to
fruit drop, the potential for total seed destruction by multiple
bruchid generations within a single acacia crown would be very high.
The bruchid larvae bore through the pod wall, into the seeds, and
mature one to a seed. Presumably there is intense intra- and inter-
specific competition among the two species of bruchid, as there can
easily be ten times as many eggs laid on a pod as there are seeds
inside. Presumably, as in other bruchids (e.g., *Scheelea* palm nut
bruchids, Janzen 1971b), the larger larva simply eats the smaller

invaders. About a month later, the adult *Mimosestes* emerges through the side of the seed, cuts a hole through the pod wall (or uses some other bruchid's exit hole), and emerges. Apparently, immediately after mating, the female can oviposit within the pod crop. This suggests strong selection against asynchronous pod crops within an individual's crown, and should select strongly for intra-population synchrony if the acacias are close enough to mutually infect each other with bruchids.

The bruchids will also oviposit on pods that have fallen to the ground, which brings out an interesting aspect of their coevolution with the acacia. When the pods are on the acacia, the beetles lay their eggs on all sides of the pod; presumably when a small object like an acacia pod is fully exposed to the breeze, no side is a substantially better microclimate than another. However, once the pod is lying on the fully insolated ground, the situation changes dramatically. There is almost no air movement right at the ground surface, and the soil surface attains temperatures of 50 to 70°C. The female, who oviposits at night, dusk and dawn, responds by laying her eggs almost entirely on the underside of the pod, which is presumably cooler and may be slightly more humid as it is directly against the ground. For example, 25 pods on the ground under an *A. farnesiana* tree in the dense clump had the following number of eggs on the surface of the pods facing upward: 2, 4, 3, 4, 2, 3, 1, 0, 6, 3, 2, 6, 4, 2, 5, 6, 2, 5, 6, 4, 1, 4, 0, 1, and 0 ($\overline{X}$ = 3.04); on the surface of the pods facing downward, the same pods had 27, 12, 38, 33, 13, 29, 23, 12, 15, 5, 18, 17, 38, 15, 18, 17, 8, 13, 32, 10, 34, 23, 37, 4, and 9 eggs ($\overline{X}$ = 20.0). The eggs on the upper surface were probably laid on the pods before they fell. These numbers of eggs per pod are by no means exceptional for pods on the ground. The number of eggs per pod on the tree averages about 8, with most values falling between 1 and 15. However, though virtually every egg laid on a pod on the tree produces a larva that bores down through the maturing pod wall and into the relatively soft seed, the larvae from eggs on the fallen mature pods have a difficult time penetrating the hard pod wall and the hard seed coat. Dissections of dry hard pods that have been oviposited on *in vitro* showed that once the pod has been off the tree and dried for a month, the larvae do not make it from the egg to the inside of the seed. They can, however, mature from eggs laid directly on mature seeds in the laboratory. It is not known if either species of *Mimosestes* is more responsible for the eggs on the fallen pods or on the pods still on the tree.

If the bruchids were to lay these eggs on the pods while on the tree, at least half of them that were laid near the time of pod fall would run the chance of being killed, since at least half of them would on the average end up on the upperside of the fallen pod. Such a situation should select for a bimodality of oviposition behavior. The female should oviposit indiscriminantly on the pod surface until about the time that the pod is ready to fall, and then stop until the pod is on the ground. It is possible that each species of *Mimosestes* occupies one of these peaks.

With this background in hand we are ready to examine the outcome of asking "do the bruchids kill more *A. farnesiana* seeds in the dense clump than in the diffuse clump"? The first attempt at answering this question, and the only one to be discussed here, was to simply collect up to 200 pods from the ground below each of 36 *A. farnesiana* in the dense clump and 11 in the diffuse clump. The idea was to mimic a dispersal agent removing a set of pods, with the intent of thereby being able to state how many seeds survive bruchid attack (this assumes that these bruchids do not locate dispersed pods or seeds). The method, and interpretation of results, suffer from not knowing how closely the sampling time matches the time at which dispersal agents would normally have removed seeds or pods. The pods were placed in plastic bags and the beetles allowed to emerge. They were shipped to the University of Chicago and a sample of 30 to 100 pods from each bag was x-rayed within six weeks of collection in order to determine how many seeds were intact, aborted, contained bruchid immatures or had exit holes from them (mammography x-ray film is ideal for this purpose). There were no signs of small bruchid larvae in the x-rayed seeds, indicating that in spite of re-oviposition on pods in the bags, none of the larvae had managed to re-infest the seeds.

The results of this sample are presented in Tables 1 and 2. The raw data are presented in full since such data are non-existent in the literature and I suspect may be of use to later workers as our understanding of the interaction between bruchids and seeds increases. There are a number of glaring conclusions to be drawn, but their interpretation is far from simple. In short, in the dense stand of *A. farnesiana*, there were significantly greater numbers of pods per tree and seeds per pod; the dense stand also had an apparent lower percent seed mortality by bruchids and an apparent higher percent of aborted seeds (cf. Tables 1 and 2). However, if these are added, we find that the percent dead seeds is identical in both samples. In the final summation, then, the *A. farnesiana* in the dense clump produced nearly 15 times as many viable seeds per tree as did those in the diffuse stand, but it was done by producing 7 times as many pods per tree as in the diffuse stand rather than through differential seed mortality in the two stands. However, as a member of my audience once said to me long ago, the easy thing is rejecting the null hypothesis. The hard thing is figuring out what alternative to accept. I think that it will be profitable to take each of these parameters and examine it in turn in the ecology of the acacia and the bruchid.

Table 1. Statistics on pod crops of *Acacia farnesiana* in the dense stand (n = 36). For comparison with pod crops in the diffuse stand see Table 2 (see text for details).

Number of pods on tree (est.)	$\overline{X}$ of seeds/pod	% seeds killed by bruchids	% seeds "aborted"	Number of viable seeds produced by tree (est.)
1700	9.6	7.78	88.52	604
340	10.2	10.35	86.10	123
850	8.8	12.03	79.04	668
325	9.7	12.17	85.93	60
669	9.1	13.67	73.83	761
1200	11.8	15.76	77.41	967
1740	10.8	17.34	78.32	816
83	8.1	20.80	72.12	47
5020	9.5	21.94	71.73	3019
350	10.1	28.03	58.92	461
1150	9.8	28.42	62.81	988
57	10.1	28.67	66.33	29
183	14.3	29.52	66.98	91
1175	8.8	29.93	67.15	302
370	9.1	31.18	68.44	13
1622	12.1	35.92	7.08	11186
483	9.9	40.94	55.43	173
1600	6.1	41.77	45.57	1235
1800	4.0	44.90	46.94	587
1800	12.0	51.91	6.08	9073
25	11.1	52.25	27.03	59
240	10.2	52.91	24.45	554
13000	10.5	53.21	12.65	55364
481	10.3	53.35	35.67	544
800	11.1	54.99	10.47	3067
750	9.4	56.35	11.09	2295
6250	9.9	61.38	5.78	20319
1020	10.0	63.93	15.43	2105
750	10.8	64.05	5.48	2468
4600	11.4	64.20	5.34	15974
45	8.7	67.69	5.38	105
575	10.6	69.58	22.73	468
1400	10.2	72.00	5.80	3170
1002	12.9	73.10	3.14	3071
50	7.3	73.59	14.44	43
7500	12.2	74.30	6.82	17276
$\overline{X}=$ 1695	10.01	41.30[†]	38.40[†]	4391
S.D.=2680	1.90	23.90[†]	35.41[†]	10444

[†]Statistics based on arc sin angular transformation.

Table 2. Statistics on pod crops of *Acacia farnesiana* in the diffuse stand (n = 11). For comparison with pod crops in the dense stand see Table 1 (see text for details). t values for differences between dense and diffuse stands are provided, all of which are significant.

Number of pods on tree (est.)	$\overline{X}$ of seeds/pod	% seeds killed by bruchids	% seeds "aborted"	Number of viable seeds produced by tree (est.)
8	7.1	46.00	50.00	31
29	8.4	46.21	31.15	131
300	9.6	50.47	1.89	1427
20	8.6	53.33	5.00	80
23	7.2	55.41	21.23	74
270	9.0	56.96	42.62	1174
200	7.8	58.10	37.71	654
82	11.2	61.88	18.06	350
26	7.4	65.10	30.73	67
91	7.1	69.76	15.73	195
1600	6.2	76.77	13.03	2304
$\overline{X}$= 240	8.25	58.35[†]	21.23[†]	301
S.D.= 463	1.51	11.56[†]	18.65[†]	540
t_{45df}= 3.1090**	2.872***	3.4049***	2.1714*	2.3335*

[†]Statistics based on arc sin angular transformation.

Numbers of Pods per Tree

There are two distinctly different potential causes for the greater numbers of pods on the clumped trees. First, the clumped trees could be growing on a site in which *A. farnesiana* adults can harvest more resources than in the diffuse area. While the crowns of the *A. farnesiana* in the diffuse area were not noticeably more crowded by themselves or other species than in the clumped area, one can say nothing of root competition, soil nutrients, soil drainage regimes, drought impacts, etc. One could even argue that the rarity of adults in the diffuse area is a reflection of this, operating through the difficulty of seed becoming established in that micro-site. However, such a conclusion is confounded by the fact that the adults in the clumped site are producing nearly 15 times as many viable seeds as are the adults in the diffuse site (though it is clear that this need not necessarily lead to a higher density of adults). Such a "resource hypothesis" would be very hard to test unless it happens to be that the nutrients in short supply could be replaced by fertilizers.

Second, the widely spaced trees could be directly pollinator-limited. I am reluctant to accept this as a general reason for small seed crops, as it appears that a tree species would soon adjust its degree of selfing to pollinator ability. However, it can obviously be a cause of small seed crops for small fractions of the population, especially if they carry the genetic material possessed by the majority of the population, a majority that is not normally pollinator-limited. In the case discussed here, it is quite possible that the dense clump, when in flower, produced a very large pollen resource, one in which it was more profitable to forage than the scattered *A. farnesiana* a short distance away.

The third possibility is that dispersal agents are removing the pods much more rapidly from beneath the widely spaced *A. farnesiana*. This possibility was appreciated at the time, and I searched the acacia crowns to see if the numbers of infructescence stalks in the crown was in approximate agreement with the numbers of pods on the ground. It was, so this possibility must be rejected.

With the data at hand, there is no way of choosing between one of the first two hypotheses. This is so even if the inflorescence scars can still be counted. If the scattered trees are genetically programmed for life in the habitat containing the high concentration of *A. farnesiana*, they may well bear flowers in numbers appropriate to the pollinators that normally frequent such dense stands, and may produce flowers in numbers appropriate to the size of seed crop that can be produced by the energy reserves of a plant growing in such a site. However, just as mentioned above for the resource hypothesis, the pollinator hypothesis can be easily tested. One simply has to hand-pollinate the flowers of the scattered plants and then examine the size of the viable seed crop.

There is, however, an even more confusing alternative available. If the widely scattered plants are "programmed for a scattered life", with their flower crops adjusted to both the numbers of pollinators that normally come to the plants, and to the amounts of reserves normally gathered by such plants, then hand-pollination could give a quite false impression of what is going on. The plants might well set more seed and then come to vegetative disaster during the following dry season, or in the next severe crown competition to which they were subjected. I would have concluded that they were pollinator-limited, when in fact they are energy-limited but have not evolved a mechanism to "know" when too many pods have been set and depend on the low pollinator activity to determine this. However, I personally doubt that wild plants become quite so dependent on external factors with such high a potential variance as the number of bees to arrive at a flowering tree from year to year. I should at this time point out that no one has ever recorded this parameter for a tropical plant (and I have never seen such information for a mid-latitude plant). By the same token, it seems highly unlikely that a tree would rely on having the same amount of resources accumulate each year with which to make fruits. Perennial plants must therefore have some internal mechanism to decide how many pods can be set with the reserves of a given year, and they must do it early in the progression of flower primordia to bud to flower to immature pod.

Numbers of Seeds per Pod

This is a most perplexing parameter. Numerous selective forces potentially affect the phenotypic trait "numbers of bits into which the total seed crop is divided".

As the number of seeds per pod declines, the amount of resources (pod walls, pod sugars, pod pulp) expended per seed increases. I hasten to add however, that the amount "spent" per seed *dispersed* need not necessarily increase. This depends entirely on how the dispersal agent community as a whole responds to small packages, large packages, or some size distribution of packages. As the actual animals (how many of what species) to arrive at a given tree depend on that tree's exact location, the tree genotype can be molded only toward some sort of "average" pod crop and dispersal. The variance in seeds per pod that we see among the *A. farnesiana* in each of the sites may be the outcome of direct selection for variance, or the outcome of various selective factors tugging in different directions, or some more horrendous combination of the two. The differences between the two sites can thus be a result of active selection, if each of the two habitats represents a distinct deme. I think this is not likely, though genetic differences between such proximate portions of a plant species are certainly well known. On the other hand, the difference in mean numbers of seeds per pod could also be simply the result of the two factors discussed in the previous section, to wit, either the trees are starving or not enough pollen grains are arriving on the stigmas.

Unfortunately, we have no idea if tropical trees vary the numbers of seeds per pod as the resources available to the tree vary. However, it is quite conceivable that the ability of the tree to attract dispersal agents changes as the pod crop size declines, and with a different array of dispersal agents arriving, the optimal number of seeds per pod should be different. It should be noted that the number of seeds per pod declines in two ways with *A. farnesiana*. On the one hand, there may be fewer seeds in a pod of the same apparent size, and on the other hand, there may be smaller (shorter) pods. Both cases occurred in the samples at hand, but such a distinction was not made in the original census. It was quite obvious, however, that certain trees produced almost entirely very short pods, and other trees produced almost entirely very long pods; this difference in pod length is largely responsible for the differences in mean numbers of seeds per pod from tree to tree, and this in turn means that the widely spaced *A. farnesiana* trees more often had short pods than did the *A. farnesiana* in the dense clump. This conclusion, incidentally, implies that it is reserves that are in short supply rather than pollinators, because if it is only pollinators, there would be no reason for those flowers that *were* pollinated to produce small pods. Furthermore, if it were the number of pollen grains arriving at the stigma that was setting the numbers of seeds at a lower level, then we would expect a high variance in the sizes of pods within a crop, rather than for some trees to have many short pods and other trees to have many long pods.

The bruchids have to be considered as part of the selection for numbers of seeds per pod as well. They figure in two ways. First, if fruit size is important in the *rate* at which seeds are removed from the tree, then a factor in selection for an optimal fruit size will be that the longer the pods stay at the tree, the more seeds will be killed by the bruchids. Second, it is quite possible that there is an optimal distribution of pod sizes (or seeds among the pods) that minimizes the number of seeds actually gotten by the bruchids before dispersal.

Forgetting the dispersal agents and developmental impossibilities for a moment, we may imagine at one extreme the seeds contained in a small number of 1 to 2 meter long pods. At the other extreme, each infructescence could be festooned with 2 cm long pods, each containing only a couple of seeds. If the ends of the pods were done up economically, the many-pod morph might cost little more than the long-pod morph. There is no information that would allow me to determine which of these two extremes might lose the most seeds to *Mimosestes sallaei* and *M. immunis*. These two beetles lay their single eggs at short intervals along the pods and there is no obvious way that one pod structure would result in fewer bruchid-killed seeds than the other. If the pod were long, then the bruchid could simply walk along it laying eggs. When the pods are divided into many small pieces, the bruchid could walk down one, then to the next, and so on.

It is possible that there are external factors operating. A bruchid on the two different pod types could be exposed quite differentially to predators. I cannot, however, see how one of these two contrasting pod types could favor hymenopteran parasitoids (to say nothing of the fact that I cannot recognize how it would be selected for even if it did, provided that the bruchid killed the seed before the parasitoid killed the bruchid).

Finally, we must add that the physiological and developmental costs of the two extreme types of pods should be considered. There are numerous legume pods with many more seeds per pod than that of *A. farnesiana*, and likewise, many with far fewer. This does not free us from the problem that as the selection pressure for a larger number of seeds per pod slides upward, the physiological cost of responding to it also increases, through such things as having to increase the average number of pollen grains (or the size of the polyad) that hit the stigma, which may in turn require changes in the floral attraction to attract a different set of pollinators, etc.

Percent Seeds Dying

This apparently simple parameter has turned out to be the most confusing of the lot. The per cent aborted seeds is significantly lower in the widely scattered acacias but the percent aborted seeds is significantly lower in the widely scattered acacias but the percent unambiguously bruchid-killed seeds is significantly greater to a degree such as to exactly complement the abortion percentage; in the dense acacia site, 21.30% of the seeds were viable and in the diffuse acacia site, 21.35% of the seeds were viable.

The most attractive hypothesis is that the plant aborts a seed when a bruchid larva enters it, if seed development has not progressed past a certain point. If this hypothesis represents the real world, then it suggests that either the pods in the diffuse stand mature earlier or that the bruchids find the pods later in their maturation cycle. Whichever the case, the bruchids kill the same number of seeds in the end in both stands. This simply suggests that from the bruchid's standpoint, the acacias in the diffuse stand are not far enough apart to cause a reduction in seed predation, *at the smaller sizes of seed crops found in the diffuse stand.* If the average seed crop per acacia in the diffuse stand had been the same as that in the dense clump, then it seems reasonable that there might have been a lower percent mortality in the diffuse stand. This assumes that a single tree with a large pod crop is not as conspicuous (in its odor) as is a clump of A. *farnesiana* in fruit.

A second reasonable hypothesis is that the bruchids may search out pods or trees with a low number of aborted seeds. This appears less likely, but should not be discounted. If this should turn out to be the case, then it suggests that the very large number of pods present in the dense clump habitat have satiated the bruchids' ovipositional ability to some degree. It also requires an explanation for why the physiological abortion rate should be lower in the widely spaced plants. A lower physiological abortion rate in the widely spaced acacias would favor the idea that they are pollinator-limited rather than resource-limited.

CARRYING CAPACITY

It is tempting to intuitively consider the number of bruchids
generated by the host population's seed crop each season as a
primary parameter in determining how many bruchids arrive at the
seed crop in the next season. This, however, assumes that the adult
mortality between seed crops takes some fairly consistent form each
year, and may be practically considered to be independent of host
and bruchid density. The latter assumption requires that any one
species of adult bruchid makes up such a small portion of any general
predator's diet that even if the beetles were to be abnormally
common, it is unlikely that a generalist would temporarily specialize
on them. Furthermore, there do not appear to be any specialist
(arthropod or vertebrate) predators on adult bruchids moving in the
habitat at large. However, there are two things wrong with such a
simplistic view. First, the adult bruchids (at least some species)
do take nectar and pollen from hosts and this may be a resource
for which they compete as adults (nectar and pollen fed to laboratory
bruchids increases both longevity and fecundity). Second, the adult
beetles may congregate in particularly desirable sites in the forest
and these high density points may attract predators.

Resource Size of Adult Host Plants

Mimosa pigra (Mimosaceae) is a common low shrub in marshy cattle
pastures of Guanacaste (and much of lowland Mexico and Central America), where its seeds are attacked by the larvae of *Acanthoscelides
pigrae*, a small bruchid host-specific to *M. pigra*. The natural
habitat of this plant is marshy areas along frequently flooding
rivers and on river banks. At Palo Verde, mature pods may be
found on the plant from late July through March, and the first
flowers appear at the beginning of the rainy season (April-May).
A new crop of flowers is produced each day. The adult bruchids may
be found on the flowers at dawn, appearing to feed on the pollen.

To examine how many adult beetles are associated with a flowering plant, an isolated *M. pigra* bush was selected at the south end
of the pasture immediately to the south of the Palo Verde Field
Station mentioned previously. The nearest other *M. pigra* bushes
were 212 meters away, and these bushes formed a dense patch of
about one half hectare in area (i.e., a very large potential source
of *A. pigrae*). On the mornings of July 10 to 12, in 1971, I
collected all the beetles on all the flowers, for totals of 312,
289, and 314 adults. On the mornings of July 13 to 16, I caught
only 26, 19, 24, and 25 beetles. On the mornings of July 17 to
22, I caught the beetles but released them afterward at the center
of the bush. The numbers were 24, 41, 72, 69, 136, and 105. It
rained on the nights of July 19-20 and 21-22, the two nights

preceding mornings in which the beetle numbers did not increase.
I interpret these results to mean that the shrub had a carrying
capacity of about 300 beetles, and the "island" had an immigration
rate of about 20 to 30 adult beetles per night, except when it
rained. (This plant would produce a minimum of 60,000 seeds during
this fruiting season.)

It seems evident that in this example, conditions were crowded
enough somewhere, a good two months after the last seed crop, for
the beetles to be migrating among the plants that were to be their
oviposition hosts as well as the adult food hosts. In migrating,
they expose themselves to a wide variety of mortality factors which
can in this sense be viewed as density-dependent. Furthermore,
once they arrive at a bush, it appears that the carrying capacity
of the inflorescences is very substantially less than that of the
bush as a whole, again indicating that conditions may be crowded
on the inflorescences.

It should be noted here, however, that *M. pigra* is exceptional
among the bruchid larval hosts in the deciduous forest in that the
adults are abundant on the flowers. Adults are occasionally taken
in sweep samples of general vegetation many hundreds of meters from
M. pigra plants, but they have not been found on the flowers of
any other species of plant.

Concentrations of Adults

As reported in a recent sweep sample study of seasonal changes
in insect density (Janzen 1975a), large numbers of small beetles
may become concentrated on the shaded understory foliage beneath
individual evergreen trees in the deciduous forest during the last
half of the dry season. A substantial number of these beetles are
bruchids and weevils that probably emerged from seeds earlier in
the dry season. It is conceivable that an exceptionally large
number of bruchids of one species could contribute to causing
small insectivorous birds to concentrate their foraging activity
at these sites. However, if this is the case, it is also possible
for an outbreak of any other species of insect to do the same, which
may be one of the multitude of ways in which the upper limits of
insect species richness in such a habitat are set.

AVOID THY NEIGHBOR

A core aspect of the theory advanced earlier (Janzen 1970) is
that scattered or dispersed seeds are less likely to be found by
seed predators than are seeds in the cluster represented by an undispersed seed crop (either still hanging in the tree or merely
fallen to the ground below). That such dispersal is effective has

been shown with seedlings of a deciduous forest legume (*Dioclea megacarpa*, Janzen 1971c), nuts of a large palm (*Scheelea rostrata*, Wilson and Janzen 1972) and seeds of a sterculiaceous tree (*Sterculia apetala*, Janzen 1972). However, it should be recognized that, just as with the dense and diffuse clumps of *Acacia farnesiana*, there are lower limits to the distances that seeds can be separated and show a reduction in seed predation. An example is offered by *Spondias mombin* (Anacardiaceae).

S. mombin is a common tree in the deciduous forest around the Palo Verde Field Station. It produces large numbers of ripe fruits from the middle to the end of the dry season, fruits that are eaten entire by monkeys (and probably by birds and other mammals). The large multi-seeded woody nuts are then defecated onto the forest floor somewhat cleaned of their outer pulp. At this time, a large undescribed species of *Amblycerus* bruchid searches them out and glues a single egg on each one found. The larva bores in and eats all the seeds. However, with the present density of dispersal agents in this forest, the majority of the fruits end up rotting on the ground below the parent tree. Once the fruit has rotted off, the *Amblycerus* oviposits on them just as if the fruit had been peeled off by a dispersal agent. This concentration of nuts beneath the parent may be viewed as either a "bruchid sink" that lowers the number of beetles that find the dispersed nuts, or as a "bruchid generator" that probably provides 90% or better of each new generation of bruchids. Incidentally, this is another way that the presence of seed dispersal agents can clearly affect the density of the seed predator, which may in turn influence the density of adult trees.

To examine the intensity of seed predation directly below the crown of the *S. mombin* trees and compare it with that at the outer edge of the seed shadow produced only by fallen nuts, about 100, 1-2 month old nuts were collected from each position from beneath 12 large *S. mombin* growing in the forest about 4 km northwest of the Palo Verde Field Station. The density of nuts can be as high as 200 in a square meter, directly beneath the crown, but the average is about 20 per square meter. The percent of nuts with *Amblycerus* exit holes at the base of the tree and directly under the outer edge of the crown (4 to 8 meters from the base of the tree) were respectively 26, 26, 22, 16, 11, 48, 44, 8, 23, 61, 14, and 38 ($\overline{X}$ = 28), and 35, 19, 28, 10, 5, 45, 46, 13, 20, 57, 12, and 40 ($\overline{X}$ = 28). There is no hint of a difference between the two means. Furthermore, if the comparison is made pair-wise, the number of exits from nuts at the base of the tree agrees closely with the number of exits from nuts under the outer margin of the crown. A careful dissection of 523 nuts from below three *S. mombin* show that 26% of the nuts lacking exit holes but with bruchid eggs on them have their seeds killed by a bruchid, but the bruchid died before it could emerge. The remaining 74% of the nuts with eggs

on them had aborted or rotted seeds inside (cause unknown, but could
well be that the bruchid larva attacks the seed and then dies while
too small to be seen later on in the decomposing material, as happens
with bruchids in *Scheelea* palm seeds). If we add the nuts with eggs
on them to the nuts with exit holes, as an absolute measure of how
many nuts were at least found by an *Amblycerus* bruchid, the respective percentages are 36, 50, 55, 60, 60, 66, 50, 27, 46, 71, 36,
and 45 ($\overline{X}$ = 50) for the nuts at the base of the tree, and 41, 49,
49, 58, 19, 69, 46, 43, 30, 61, 26, and 46 ($\overline{X}$ = 45) for the nuts
under the outer edge of the crown. The differences in the means
are not significant. To understand why the remaining 50 percent
of the seeds did not generate a large seedling shadow beneath the
parent trees, I should point out that of the 523 nuts mentioned
above, only 4 contained a viable seed (4% of the 102 seeds that had
neither *Amblycerus* exits nor eggs on the outside).

HOST SPECIFICITY

In the Guanacaste deciduous forest, the insects that breed in
seeds show amazing specificity. I should add that I have seen nothing in other tropical vegetation to suggest that this is a phenomenon
peculiar to Guanacaste deciduous forest. Detailed documentation of
this specificity is still in progress, but it seems reasonable to
present a preliminary view here as there is no indication that com-
pletion of the study will drastically change the picture (the data
given here are a refinement on the first approximations mentioned
in Janzen 1973a,b; 1974b).

In a flora with about 300 to 400 species of plants (excluding
grasses) with seeds large enough for a bruchid or other seed-eating
insect to develop in, I can state with certainty that at least 69
species have no pre-dispersal seed predation by insects, 69 species
have one or more species of bruchids, 6 have weevils, 2 have cerambycids, and 5 have moth larvae (excluding seed chalcids). In addition,
at least 20 other species of bruchids have been collected in and
around the forest but their hosts are unknown. This means that a
minimum of about 25 to 33% of all the plant species in the deciduous
forest suffer considerably from pre-dispersal seed predation by
insects that develop in the seeds. If we consider only the large
woody plants, this figure moves upward to between 50 and 75%. There
is one major source of pre-dispersal seed predation by insects that
is not considered here; bugs (Hemiptera) take a very large number of
developing embryos of certain species (these are often recorded as
aborted seeds) and weevils (Curculionidae) develop in flower buds
of many species. We can ask two questions of this information.
How are the bruchids distributed among the plants, and how are the
plants distributed among the bruchids?

Among the 88 bruchid species whose hosts are known with

certainty to date, 73 (83%) have only one host species. All the
weevils and cerambycids have only one host species. Thirteen species
of bruchids (15%) have 2 hosts (e.g., *Gibbobruchus guanacaste* in
Bauhinia pauletia and *B. ungulata*, *Amblycerus championi* in *Cordia
dentata* and *C. panamensis*, *Acanthoscelides kingsolveri* in two species
of *Indigofera*, *Ctenocolum tuberculatum* in *Lonchocarpus costaricensis*
and *L. minimiflorus*, and *Megacerus leucospilus* in *Ipomoea pes-caprae*
and *I. fistulosa*). One bruchid has three hosts (*Ctenocolum crotonae*
in *Lonchocarpus costaricensis*, *Lonchocarpus minimiflorus*, and
Piscidia carthagenensis) and one has eight and perhaps more (*Stator
limbatus* in *Pithecellobium saman*, *Pithecellobium dulce*, two species
of *Albizia*, two species of *Lysiloma*, and two species of *Acacia*).
All indications are that location of the hosts of the twenty or so
un-reared species will swell the "one host only" category.

There are undoubtedly a number of components to the answer to
why these seed predators are so host-specific. Certainly it is not
the mere outcome of segregation by habitat; with careful collecting,
as many as 40 species of bruchids may be reared from the seeds in a
five hectare plot, and 22 species were taken in one forest understory
sweep sample of 800 sweeps (dry season, Palo Verde, study reported
in Janzen 1975a and below). On the other hand, in over half of the
cases where a bruchid occurs in two species, the two species of
plants are quite definitely habitat-segregated. A conspicuous
example is *Megacerus leucospilus* in the seeds of *Ipomoea pes-caprae*
growing on ocean dunes and in the seeds of *Ipomoea fistulosa* growing
in freshwater marshes along the Rio Tempisque.

Competition among bruchids within a species of plant may contribute to the host specificity. If we regard those species with
bruchids as in some sense available, the bruchids appear to be quite
evenly distributed among them; of 69 host species, 55% have one
bruchid and 35% have two bruchids, while only 7% have 3 bruchids and
3% have 4 bruchids. This concentration of 90% of the plants in the
1 or 2 bruchid range implies that if the numbers rise over two
species, inter-bruchid species competition may become very severe.
In all cases of 3 or 4 bruchids per host plant species, at least 90%
of the bruchids reared from seed samples belong to two species.
Such competition may be selecting in favor of specialists by selecting for maximal efficiency on a given host.

The most obvious candidate for a major cause of the specialization is the need to specialize to overcome the chemical, behavioral
and morphological peculiarities of the host plant. In short, one is
forced to postulate that as the plant becomes better defended
(selected for by past generations of bruchids), the beetle has to be
progressively more specialized to stay with its host. The more
specialized it is on one host, the more difficult it should then be
for it to also be adequately specialized to attack a second host.
Such incompatibility may take two forms. (1) The critical traits

of the attacker of two host species may be chemically difficult to possess simultaneously. For example, the enzyme system that breaks down the toxic secondary compounds in one species' seeds may be bio-chemically incompatible with the enzyme system necessary to break down the secondary compounds in the seeds of the second host species. Again, to locate the pods of one species, the female bruchid might have to search in moist and dark micro-habitats (e.g., *Spondias mombin* nuts being sought by *Amblycerus* on the forest floor) and in sunny and dry habitats for the other (e.g., *Combretum farinosum* fruits being sought by *Amblycerus* high in a deciduous forest canopy). (2) The critical traits of the attacker of two host species may be energetically (or nutritionally) difficult to maintain simultaneously. For example, it may be energetically very difficult for *Caryedes brasiliensis* to have the enzyme system to both deal with seeds with 5 to 10% canavanine in them (which it does when feeding on *Dioclea megacarpa*, Janzen 1971c) and seeds with a 5 to 10% concentration of L-DOPA as found in *Mucuna andreana* (Bell and Janzen 1971), which grows in nearby habitats and has pods and seeds similar to those of *D. megacarpa*.

In the previous paragraph I was careful not to imply that any combination of attacking abilities is impossible in any absolute sense, as it is clearly not. *Stator limbatus* is the most amazing in this respect. It is clearly a specialist at laying its eggs directly on the exposed but undispersed seeds of four genera of mimosaceous legumes. Associated with this, however, it must be able to deal with the alkaloid pithecolobine in *Pithecellobium saman* seeds (Magnus and Seaforth 1965), the uncommon amino acid albizziine in *Albizia* seeds, and probably other secondary compounds in *Lysiloma* and *Acacia* seeds. In all the other cases of 2 or 3 hosts being attacked by one bruchid, the females oviposit on the intact fruit (and the larvae bore through the fruit wall to the seeds inside), and in all cases the fruits of the two or three hosts are very similar in shape, texture, thickness, size, and odor. The pair of hosts are also congeneric in all cases but one (*Lonchocarpus* and *Piscidia*). Center and Johnson (1974) have recently stressed the fruit wall as a major barrier to bruchid entry, and these results support that emphasis.

There are five ways to get at the role of seed chemistry in this extreme host specificity.

(1) We can ask how readily the bruchids transfer onto intro-duced species of plants. I have only one example to contribute, that of the shrub *Cassia alata* (which may be native to Central America, but is certainly not wild in the Guanacaste deciduous forest). *C. alata* in gardens very rarely have an *Amblycerus* and a *Sennius* in their seeds, bruchid species that normally live in one or two species of herbaceous weedy *Cassia* that have pods very different from those of *Cassia alata* in superficial aspect. Since

the native and introduced *Cassia* probably have very similar seed
chemistry, this example is of very little use. In general, however,
it can be stated that none of the numerous introduced legumes (e.g.,
Delonix regia, *Caesalpinia* sp., *Erythrina* spp., *Phaseolus vulgaris*,
Lathyrus sp.) in Guanacaste have had indigenous bruchids move onto
them.

(2) We can ask what happens when a wild female beetle accidentally oviposits on the "wrong" host. The technology of this
situation is almost impossible since the only time that a record
will be obtained is if the bruchid larva survives. (For the most
part, bruchid eggs on the fruit or seed cannot be identified with
ease.) However, in all the tens of thousands of bruchids that have
been reared to date in this study, I have only one unambiguous case
of a bruchid being reared from seeds of a wild plant other than its
usual host. At Finca Taboga in southern Guanacaste, R. Carroll found
a pile of howler monkey feces containing nuts of *Spondias mombin*
(Anacardiaceae) and the similar-sized seeds of *Eugenia salamensis*
(*Psidium rensonianum*; Myrtaceae). The site smelled strongly of
S. mombin fruits. There were 78 *S. mombin* nuts with *Amblycerus*
eggs on them and 6 *E. salamensis* seeds with one *Amblycerus* egg on
each (out of a total of 354 *E. salamensis* seeds and 197 *S. mombin*
nuts). During the following months, the *Amblycerus* emerged from 71
of the *S. mombin* nuts and a single adult emerged from each of 5 of
the *E. salamensis* seeds. A bruchid has never been reared from any
other of thousands of *E. salamensis* seeds, though an undescribed
weevil kills about 5% of large seed crops.

(3) A small amount of data has been gathered about the outcome
of offering a female bruchid a series of seeds of wild hosts from
her habitat on which to oviposit, but this method has not been
exploited fully owing to the difficulty of maintaining most wild
bruchids as laboratory colonies. *Mimosestes sallaei* females have
been found to oviposit on almost any large smooth round seed if there
are *Acacia farnesiana* seeds present in the same container. One
experiment has shown that the larvae from eggs laid in this manner
can survive in *Acacia cornigera* and *Acacia collinsii* seeds (both
occur in the same general habitat with *A. farnesiana*, both are
attacked by bruchids, but neither is attacked by *M. sallaei* in
nature). *M. sallaei* do not survive in seeds of *Enterolobium cyclocarpum* (containing albizziine), *Mucuna andreana* (containing L-DOPA),
Caesalpinia crista (=*Guilandina crista*, containing gamma-methyl-
glutamic acid), *Canavalia maritima* (containing canavanine),
Schizolobium parahybum (containing schizolobine), *Dioclea megacarpa*
(containing canavanine). Only the last mentioned species of plant
has a bruchid in its seeds, but all occur within the flight range
of *M. sallaei* and none have pod walls thicker or noticeably tougher
than those of *Acacia farnesiana*.

(4) A descriptive analysis of seed secondary compound content

over an entire habitat would probably be helpful, but not definitive. If all the seeds in a habitat were found to have similar or the same secondary compounds in them, it would be unlikely that the bruchids would be specific on account of seed chemistry. This analysis is still underway, but it is clear that the diversity of secondary compounds is at the level of about 5 different major constituents for each 10 species of seeds. In addition to the major compounds occurring at 3 to 10% concentration, there are often others that occur at lower concentrations, to say nothing of the concentration of lectins, heteropolysaccharides, endopeptidases, and saponins, all of which occur in bruchid host (and non-host) seeds, are toxic to certain bruchids, and have not yet been extensively sought in seeds of Guanacaste legumes.

(5) Secondary compounds from seeds can be incorporated in bruchid artificial diets, and such a study is currently underway in collaboration with E. A. Bell (nitrogenous secondary compounds) and I. E. Liener (lectins or phytohaemagglutinins). I chose *Callosobruchus maculatus* (stock obtained from the USDA Stored Products Insect Research and Development Laboratory, Dr. E. Jay) since this bruchid is easily reared in the laboratory and feeds readily on black-eye pea seeds (*Vigna unguiculata*) which have no alkaloids or uncommon amino acids. This bruchid may be viewed as biochemically/physiologically naive, and therefore a test of putative seed toxins on its larvae may represent what happens when a wild bruchid lays its eggs on the seeds of a non-host.

The logistics of testing are very straightforward. A Wiley mill in the University of Michigan College of Pharmacy is used to reduce dry black-eyed peas to fine flour. For control seeds, this flour is used to hand make cylindrical tablets 13 mm in diameter and 7 mm deep with a mechanical pill press (also from the College of Pharmacy). The females oviposit readily on these pills when placed in jars containing stock *C. maculatus* cultures (they also oviposit on just about any other smooth surface). All eggs but ten are removed from the pill, so as to adjust the number of developing larvae to where there will be very little intra-specific competition. An average of five to six adults emerge from the ten eggs on such a pill. For the experimental pills, shortly before the pill is made, the secondary compound is mixed in a dry powder form with the flour. To date, experiments have been with 0.1, 1, and 5% concentration of secondary compounds, as this is the range of concentrations at which secondary compounds normally occur in seeds in nature. The pills plus eggs are cultured at 80% RH at room temperature; such care is not necessary for intact seeds but the pills dry out much more rapidly than do intact seeds.

The results to date are incomplete in coverage of the secondary compounds found in Guanacaste bruchid hosts and non-hosts, but provide some interesting previews. For example, lectins or

phytohaemagglutinins, compounds well known to be toxic to mammals
when taken orally (as well as intravenously) (Liener 1974) have
been shown to be toxic to an insect seed predator for the first time.
When black bean lectin (*Phaseolus vulgaris* from Guatemala) is incorporated in the pills at 5% concentration it is lethal, and only two
beetles emerged from five pills with 1% concentration (at 0.1%,
emergence is indistinguishable from the controls).

Alkaloids incorporated in the pills may be even more lethal.
Colchicine kills all larvae at 0.1% concentration, and caffeine at
the same concentration produced only four adult beetles out of five
pills. Caffeine is lethal at 1% concentration. It is appropriate
to add here that the beetles that are produced on these marginally
toxic diets have not been checked for reproductive abilities, but
may well have reduced reproductive fitness.

Uncommon amino acids have been most extensively tested to date,
and have produced the most confusing (and interesting) results. The
following uncommon amino acids from seeds are lethal at 1% concentrations: L-DOPA, β-aminopropionitrile fumarate, L-djenkolic acid,
N-methyl tyrosine, β-cyano-L-alanine, and L-mimosine. All of these
also show very severe reduction in emergence at 0.1% concentrations.
A moderate reduction in bruchid emergence was obtained with 1% concentrations of canavanine, albizziine, D,L-2, 4-diaminobutyric acid,
γ-methylglutamic acid, and m-carboxyphenylalanine. These and others
are being tested at the 5% concentration, and canavanine, γ-methylglutamic acid, and m-carboxyphenylalanine are lethal at these concentrations. However, it should be added that D,L-pipecolic acid,
L-homoarginine, and S-carboxyethylcysteine at 1% concentrations had
no visible effect on the bruchids. In short, the effects are highly
variable, concentration dependent, but largely toxic at some level
representative of that found in seeds.

A second form of control, especially appropriate to the question
of whether uncommon amino acids in seeds are merely nitrogen storage
compounds, or defense mechanisms, or both (see Rosenthal 1972 for
an example of both in *Canavalia* seeds with canavanine in them), is
to see what effect protein amino acids have on *Callosobruchus maculatus*. At the 5% concentration, we have found the L-isomers of
tyrosine, tryptophan hydroxyproline, aspartic acid, cystine, and
methionine to be lethal, the L-isomers of cysteine, leucine, iso-
leucine and histidine to be mildly depressant, and the L-isomers
of arginine, alanine, glutamine, threonine, proline, glutamic acid,
phenylalanine, asparagine, and valine to have no obvious ill effect.
It was expected that a 5% concentration would have no effect on
C. maculatus, and so now the toxic protein amino acids are being
tried at 1% concentration. These results emphasize that simple
nutrient imbalances among species of seeds could be enough for
toxicity. It is of particular interest here that tryptophan,
cystine and methionine occur in exceptionally low concentrations

in black-eyed peas (e.g., Johnson and Raymond 1964, Evans and
Bandemer 1967, Sevilla-Eusebio et al. 1968).

HYPERPARASITOIDS

Any discussion of the coevolution of parasitoids and their
hosts, or predators and their prey must take into account their
hyperparasitoids or parasites. If we view the bruchids in the seeds
as parasitoids, then the parasitoids that they have may be viewed
as hyperparasitoids. In the Guanacaste dry forest, all bruchid
hyperparasitoids reared to date have been Hymenoptera. However,
just as one does not expect all arthropod parasitoids to have
hyperparasitoids, not all bruchids have hyperparasitoids. In
fact, none have emerged from the seed collections for at least 35%
of the 73 bruchid species mentioned earlier. From another 38%,
only one or two individual hymenopterans have been reared from
collections that produced hundreds to thousands of bruchids. Even
in the remainder, there is little suggestion that the hyperparasitoids are taking more than 10 to 20% of the bruchids (I hasten to
add, of course, that such a low percent mortality may be of great
significance in the population dynamics of the host, but it need not
necessarily be the case). It should be added here that I have not
made a conscious effort to locate egg parasites, and some undoubtedly
exist. However, in collecting pod samples in the field, egg parasites would normally be picked up along with the rest. As will be
discussed at the end of this section, low or zero levels of parasitization of tropical insects may be commonplace; the following
paragraphs are directed at why they are so low among Guanacaste
Bruchidae (there is no hint of such a phenomenon among North American
Bruchidae).

The absence of hyperparasitoids is associated in one case with
the presence of an effective disease. The two bruchids in *Scheelea
rostrata* (Palmae) seeds have no hyperparasitoids (11,000 nuts examined, Janzen 1971b) but as high as 65% of the larvae and pupae may
die in a sample of nuts. These dead immatures have their insides
reduced to a milky fluid that resembles the insides of a Japanese
beetle killed by milky spore disease. Presumably the disease is
transmitted through contamination of the female bruchid or nut
surface by bacterial spores from the litter containing rotted palm
nuts. This is the only disease of a bruchid to be found in the
Guanacaste deciduous forest (though the apparently hyperparasitoid-
free larvae of the weevils in *Andira enermis* seeds have an effective
fungus disease) and this bruchid's habitat is the wettest of all
bruchid microhabitats. I assume that the easily accessible *Scheelea*
nut bruchids (they are in the nuts for 2 to 10 months) are not
attacked by a hymenopteran hyperparasitoid because they would be
either "eaten" by the disease or indirectly outcompeted by it.
It would seem virtually impossible for the hymenopteran to outcompete

the bacterium yet allow the bruchid larva to live long enough to
cut its exit hole out of the hard palm nut.

I turn to properties of the bruchid-seed interaction to explain
the other cases of low or zero hyperparasitization of bruchids. Some
thoroughly examined cases are the following. *Cassia grandis* has
three species of bruchids in its seeds, none of which have hyperparasites in Guanacaste (though the egg clusters of *Pygiopachymerus
lineola* are very rarely attacked by an unknown hymenopteran,
Janzen 1971d). *Pithecellobium saman* has one bruchid (*Merobruchus
columbinus*) that kills 40 to 80% of its immense seed crops (a large
tree may produce 150,000 seeds annually) and no parasites have been
reared from 85 samples of 100-plus pods taken over a three year
period. The bruchids have been reared out of 23,285 fruits of
Guazuma ulmifolia (samples from 206 trees in six major habitats)
and the two species of bruchids infesting as high as 99% of these
seed crops had no hyperparasites (Janzen 1975b). Of thousands of
Amblycerus bruchids reared from *Cassia emarginata* pods, there have
been only two large Chalcidae, of a species customarily reared from
a wide variety of Lepidoptera pupae. A sample of better than 20,000
Mimosestes reared from *Caesalpinia coriaria* pods had no hyperparasitoids.

Examples such as these are probably explained by the interaction
of several factors. To be a bruchid hyperparasitoid, the hymenopteran must not only be able to survive in the bruchid larva (which
may be complicated by the presence of secondary compounds in the
larva) but it must be able to get to the larva. However, the more
species of bruchids there are in the habitat, the more species of
bruchids the hymenopteran will have to be able to attack in order
to be able to accumulate enough hosts to stay in the game. The need
for this increased generality is predicted because the more species
of bruchids there are in the habitat, the fewer individuals there
are on the average in each one of them. However, versatility is
difficult in the best of times. As the bruchids are distributed
among a very wide variety of seed chemistries, pod morphologies,
fruiting phenologies, microhabitats, seed morphologies, developmental
times, etc., it should be difficult for any one hyperparasite
species to accumulate enough attackable bruchids. The more this is
true, the more likely the community is to contain some totally
hyperparasite-free bruchids. In short, at the upper numbers of plant
and bruchid species in a habitat, the number of hyperparasite species
operating on them should decrease.

The previous paragraph can be transliterated into an even more
general case. Assume a gradient from one very common host species
to a moderate number of moderately common host species to a very
large number of quite scarce host species. Likewise assume that the
total biomass of hosts does not change over this gradient. The
number of parasitoids that can be supported along this gradient is

expected to rise first toward the middle, and then at the limit
fall off to a point considerably lower even than it started out.
I view the moderate numbers of bruchid species to be found in
southwestern North America, each with several species of hyperparasites, as representing the peak in this curve. The more distinctive is the average host species along this gradient, the more
rapidly and severely I expect the numbers of parasitoids to decline
at the upper levels of host species richness. Likewise, the more
fluctuating (predictably and unpredictably) the physical environment along this gradient, the more rapidly and severely I expect
the numbers of parasitoids to decline at the upper levels of host
species richness. Thus, for example, I expect the deciduous
tropical forest with n (large numbers) species of hosts to have as
few species of parasitoids as an adjacent evergreen tropical forest
with $2n$ species of hosts, even if the total annual production of
host biomass was the same in both sites.

Following this type of reasoning, I expect a number of Guanacaste bruchids to be very common (in most years) yet have no hyperparasitoids, simply because on some years the prey density falls to
very close to zero. In short, I am saying that the more the availability of a lower member of the food chain fluctuates the density,
the fewer specialists can survive at higher levels in that food
chain. As a case in point, there was a general drought over much
of Guanacaste at the beginning of the rainy season in 1971. In many
areas it was sufficiently severe to cause the abortion of almost all
pod crops of *Pithecellobium saman*. How *Merobruchus columbinus*, the
host-specific seed predator of *P. saman*, managed to survive is not
clear (it probably re-immigrated from neighboring areas of less
severe drought), but I can easily envisage that a host-specific
hyperparasitoid would have had an even more difficult time surviving.

The Guanacaste deciduous forest contains some other spectacular
examples of freedom from parasitoids. The very large tree *Enterolobium cyclocarpum* frequently has its entire leaf crop eaten off by
an unidentified moth larva in early June. One tree may easily
contain 20 bushels of caterpillars at a time, and tens of thousands
of them pupate under the loose bark and adjacent shelters. Neither
the larvae nor the pupae show any sign of emerging parasitoids,
though at some trees large numbers die of a disease. I have seen
at least five similar cases of other major defoliators without
parasitoids. A noteworthy characteristic of these species is that
the newly emerged adults do not return to oviposit on the new crop
of foliage produced by the defoliated or damaged tree. A tentative
hypothesis is suggested by these observations. It may be that the
herbivore is genetically programmed for a time when its hosts were
much harder to find (rather than being exposed trees poking out
of pastures as is currently the case), and total defoliation a
much rarer event. When the tree is under heavy competition in the
forest, it produces one large new leaf crop (the one that the larvae

feed on), and then turns off new leaf production even when a few
have been removed by herbivores. In the forest, the herbivore gets
one shot at the tree, builds up a moderate population of adults,
and stays in the game by being an active adult in reproductive diapause for eleven months until the next new set of leaves appears.
Such behavior appears to be adaptive in the context of the individual
female, in that she may get more eventual offspring by waiting to
oviposit rather than by exposing herself to predators while searching for a few new leaves, and by not producing caterpillars that
would be members of a relatively small overall body of arthropod
prey items which would be confronted with an array of predators
developed on the burst of prey produced by the big flush of leaves
at the beginning of the rainy season.

The hyperparasitoids that *do* survive on Guanacaste deciduous
forest bruchids are worthy of much more scrutiny, but this portion
of the study is still in an embryonic stage. Inspection of the
material reared to date along with the bruchids reveals one conspicuous fact. In strong contrast to the highly host-specific
bruchids, a given species of bruchid hyperparasitoid is found in
many species of bruchids (and I suspect, in many species of other
insects inside of seed pods and similar structures). It is of
interest to note here that it is the rare bruchids that should be
most seriously affected by such Hymenoptera. For example, the
hymenopteran might be presented with an array of 1110 pods per
hectare, which might for example contain 1000, 100, and 10 pods of
three similar legume species each of which is attacked by a specific
bruchid. If it finds 30% of the pods, a 30% reduction in absolute
numbers of the rare bruchid is much more likely to be lethal to the
population than a similar percent reduction in the common bruchid.

As mentioned at the beginning of this section, the finding that
a number of bruchids have no hyperparasitoids and that only a few
sustain several species of hymenopterans is in agreement with my
general findings with sweep samples of the Costa Rican arthropod
community. For example, in an English meadow in July, 800 sweeps
produced 225 species of hymenopteran parasitoids, which is better
than twice as many species as have been found in any Costa Rican
"old field" vegetation with 800 sweeps, and most Costa Rican samples
are about one quarter of this. The same English meadow had 2401
individuals in 800 sweeps, which is about 15 times as many as have
ever been found in a Costa Rican "old field" site. The only Costa
Rican sample that has a similar hymenopteran parasitoid species
richness to the English meadow was rainforest understory with twice
as many species of prey insects (Janzen and Pond 1975). Similar
results were obtained with an elevational transect in the Venezuelan
Andes; with increasing elevation, the hymenopteran parasitoid species
richness was proportionately the least reduced of all groups, presumably associated with the fact that the numbers of prey individuals

per species was highest at the highest elevation (Janzen et al. 1975).

SURVIVING THE INIMICAL SEASON

In the light of our mid-latitude biases, bruchids (and other seed parasitoids) in the Guanacaste deciduous forest do a very peculiar thing. Upon completing larval development, they pupate and emerge within a few weeks. In most species of hosts, there are not adequate numbers of seeds of the appropriate developmental stage for them to produce a second generation. The adults are then active, but reproductively inactive, for 9 to 11 months until the next seed crop appears. Some feed on the nectar and pollen of their host plant's flowers (e.g., *Acanthoscelides pigrae* on *Mimosa pigra* described earlier; *Acanthoscelides oblongoguttatus* takes nectar from the extra-floral nectaries of *Acacia cornigera*, its host in Veracruz, Mexico, Janzen 1967), or on nectar and pollen of other species of flowers. However, most species may be taken by general collecting with a sweep net in vegetation that is neither rich in flowers nor contains the beetle's host plants. Presumably the adults have a lower mortality rate if they actively seek out appropriately moist, cool, or shady microsites, and if they can actively avoid predators, than if they attempt to survive as a dormant individual in the damaged seed or a pupal cell in the ground or litter. In this context, it is of interest that the only Guanacaste bruchids with a conspicuous mortality from a disease are those that wait for long times in palm nuts on moist litter before emerging. Furthermore, the only seed parasitoid with a conspicuous fungal mortality are the two *Cleogonus* weevils that attack *Andira enermis* seeds; the larvae pupate in the moist soil below the seeds and may lose better than 30% of the population to an undescribed species of fungus. In short, I am saying that the seed parasitoids are not exceptions to the general hypothesis (Janzen 1973c) that a large number of tropical insects pass the inimical season as active adults, rather than as dormant individuals.

A few examples of exactly where adult bruchids are during the inimical season in the deciduous forest at Palo Verde are instructive. Adults of the *Acanthoscelides pertinax* group, *A. puellus*, *A. megacornis*, and *Stator pruininus*, and two species of *Apion* and one of *Paragoges* (Curculionidae) have been found hiding in the partially opened dried fruits of *Bixa orellana* (Bixaceae); the seeds have a thin covering of a fatty material and have *Stator championi* as their host-specific seed predator. The large clusters of dry and wind-dispersed *Alvaradoa amorphoides* (Simaroubaceae) fruits have been found to contain adults of *Zabrotes* sp., *Sennius morosus*, *Sennius* sp., two species of *Acanthoscelides*, *Paragoges* and *Phillides* (Curculionidae); *A. amorphoides* has no pre-dispersal seed predators that live in the seeds. Similar arrays of adult bruchids have been

found in the dry infructescences of *Triplaris americana* (Polygona-
ceae), another tree with no pre-dispersal seed predators in its
seeds. All three of these records are from the last half of the
dry season. A sweep sample of a stand of pure *Baltimora recta*
(Compositae) in flower in the middle of the rainy season yielded ten
males and eight females of *Caryedes quadridens*, a bruchid that breeds
in the dry season pods of *Centrosema plumieri*, a legume vine. Sweep-
ing in nearby more mixed "old field" vegetation yielded *Caryedes
quadridens*, *Amblycerus championi*, *Sennius instabilis*, and two species
of *Acanthoscelides*; the seed hosts of none of these bruchids were
available at that time (mid-July). In the adjacent deciduous forest
understory, simultaneous sweeps produced adults of *Caryedes quadridens*, *C. cavatus*, *C. x-liturus*, *Ctenocolum tuberculatum*, two species
of *Acanthoscelides*, and a species each of *Amblycerus*, *Zabrotes* and
Dahlibruchus. With the exception of the last species, whose host is
unknown, the seed hosts of none of these bruchids are available for
oviposition at that time, and only the *Amblycerus* will eventually
find its host (*Spondias mombin* seeds) within a few meters of the
vegetation swept. In the middle of the dry season (March) a sweep
sample primarily underneath scattered evergreen trees in the same
forest site yielded the following adult bruchids: *Caryedes quadridens*, *Merobruchus columbinus*, *M. solitarius*, *Gibbobruchus guanacaste*, *Ctenocolum tuberculatum*, *Megacerus impiger* group, two species
of *Zabrotes*, three species of *Stator*, two species of *Amblycerus*,
Acanthoscelides quadridentatus, *A. megacornis*, *A. pertinax* group,
A. nr. *brevipes*, *A.* sp., *Sennius instabilis*, *Sennius morosus*,
Mimosestes sallaei and *M.* sp.; the seed hosts of all of these
species are well into their infestation cycle by this time, and
oviposition has occurred long ago for most species. These adults
were undoubtedly newly emerged adults beginning their long wait
until next year's seed crop. It is noteworthy that one of the
most common large bruchids in this forest rich in trees of *Guazuma
ulmifolia* is its host-specific *Amblycerus cistelinus*. While large
numbers of these beetles were emerging from *G. ulmifolia* fruits
near the site of the dry season understory sweep samples, not one
of these beetles was taken there.

DISTURBANCE

It is fitting to close this paper with a brief discussion of
the extreme frustration of trying to understand the coevolution of
seeds and their parasitoids in habitats that have been perturbed
in various ways by exploitative western agriculture and technology
over the last 100 to 300 years (and see Janzen 1973d, 1974c).
The problem is rather straightforward. When a set of species
evolve with respect to each other for thousands of generations,
and part are then removed and the remainder have their densities
and other properties (e.g., timing of new leaf production) altered,
it becomes extremely difficult to interpret the adaptive significance

of many of the characteristics genetically programmed into the
individuals. Of course the species that are there are still inter-
acting with each other, and therefore it is quite possible to
examine their contemporary ecology, especially from a management
viewpoint. However, the generation of ecological principles, to say
nothing of evolutionary ones, is almost impossible when working with
systems that are not only not at equilibrium, but have been recently
pushed off equilibrium to an unknown degree by unknown perturbations.
For example, the pastures of Guanacaste are full of swollen-thorn
acacias containing three species of obligate acacia-ants (*Pseudo-
myrmex belti*, *P. ferruginea* and *P. nigrocincta*), all fighting actively over and coexisting on a single very narrowly defined resource;
how can one understand species-packing in such a situation when all
the physical environmental barriers that probably were once quite
important in separating these three species have been obliterated to
provide the United States with hamburger?

    The bruchid-host interaction is particularly susceptible to
the thorough types of habitat and species destruction practiced by
modern man. In trying to understand how the sizes and timing of
seed crops may have been influenced by the interaction with the
bruchid, I need to know how long the seeds of a given crop are
available to the bruchid. Except with wind-dispersed species this
is no longer possible owing to the effective elimination of all the
large dispersal agents. Even where dispersal agents are present,
the numbers of seeds they remove is closely related to what other
food sources are available, and contemporary deciduous forests with
their numerous edges, and old and new fields, are a vastly different
food resource base than a relatively undisturbed deciduous forest
with scattered indigenous farming efforts. Furthermore, as certain
dispersal agents are removed, the less desirable plants will suffer
disproportionately to the ones with the more highly desired fruits
or seeds. *Rhinochenus transversalis* and *R. stigma* (Curculionidae)
used to kill as much as 50% of the seed crop of *Hymenaea courbaril*
(Caesalpinaceae) in lowland Guanacaste; now only *R. transversalis*
occurs in most *H. courbaril* populations. *R. stigma* is going extinct
because it depends on dispersal agents to open the indehiscent pod
so it can emerge (cf. Janzen 1974c, an article in which the specific
names were accidentally reversed when giving this account). It
should be stressed that habitat alteration by man is not a new thing
(e.g., Long and Martin 1974), but it is now occurring in the tropics
at a rate and intensity far surpassing previous events.

    I also need to know how many bruchid females can be expected
to arrive at the host plant seed crop. By altering vegetation types
and changing the relative abundance of alternate adult food sources,
contemporary agriculture undoubtedly increases the numbers arriving
at some plants and species, and decreases those arriving at others.
A *Scheelea rostrata* palm growing in an open pasture will have its
seed crop almost entirely free from bruchid attack, while one growing

20 m away in forest may lose 80% of its seeds (Janzen 1971b).
However, *Scheelea rostrata* adults left growing in pastures when the
forest is cut are effectively seed-sterile, since the dense seedling
shadow is eliminated by the fierce dry season sun. Introduced species from other parts of Central America may be expected to make
this part of the story even more confusing.   H. Dingle has even
suggested that the *Dysdercus* that kills so many seeds of *Sterculia
apetala* in Guanacaste (Janzen 1972) may be an introduced species.

The effect of all of this is, of course, to cause one to seek
out somewhat less disturbed areas for study, such as Santa Rosa
National Park in northern Guanacaste and the forest near the OTS
Palo Verde Field Station (COMELCO Ranch) between Bagaces and the
Gulf of Nicoya. However, even these sites are ephemeral in the
face of irrigation schemes, inflation, population pressures, and
private interests of surrounding ranchers. The story is a familiar
one, but ecologists have a tendency to think of themselves as racing
against time. From what I see around me in the tropics, the race
is already lost.

ACKNOWLEDGEMENTS

This study has been supported by NSF GB-25189, GB-7819, and
GB-35032X, by the Organization for Tropical Studies, by the
Universities of Kansas, Chicago, and Michigan, and by the intellectual and physical input of a very large number of people working
for and with these institutions. They are mentioned by name in the
specific papers dealing with their contributions. The study would
not have been possible without the taxonomic studies of bruchids
and weevils by John M. Kingsolver and D. R. Whitehead. J. A. Chemsak,
W. D. Duckworth, S. L. Wood, R. Warner, P. A. Opler, R. Wunderlin,
W. C. Burger, and above all V. E. Rudd have contributed to deter-
minations of other groups. I. E. Liener and E. A. Bell contributed
many of the secondary compounds used in the feeding tests. Constructive criticism of the manuscript was made by D. R. Whitehead
and J. M. Kingsolver.

LITERATURE CITED

Bell, E. A., and D. H. Janzen. 1971. Medical and ecological con-
    siderations of L-DOPA and 5-HTP in seeds. Nature 229:136-137.
Center, T. D., and C. D. Johnson. 1974. Coevolution of some seed
    beetles (Coleoptera: Bruchidae) and their hosts. Ecology
    55:1096-1103.
Evans, R. J., and S. L. Bandemer. 1967. Nutritive value of legume
    seed proteins. J. Agric. Food Chem. 15:439-443.
Gwynne, M. D. 1969. The nutritive values of *Acacia* pods in relation
    to *Acacia* seed distribution by ungulates. East Afr. Wildl. J.

7:176-178.
Janzen, D. H. 1966. Coevolution of mutualism between ants and
    acacias in Central America. Evolution 20:249-275.
Janzen, D. H. 1967. Interaction of the bull's horn acacia (*Acacia
    cornigera* L.) with an ant inhabitant (*Pseudomyrmex ferruginea*
    F. Smith) in eastern Mexico. U. Kans. Sci. Bull. 47:315-558.
Janzen, D. H. 1969. Seed-eaters versus seed size, number, toxicity
    and dispersal. Evolution 23:1-27.
Janzen, D. H. 1970. Herbivores and the number of tree species in
    tropical forests. Amer. Natur. 104:501-528.
Janzen, D. H. 1971a. Seed predation by animals. Annu. Rev. Ecol.
    Syst. 2:465-492.
Janzen, D. H. 1971b. The fate of *Scheelea rostrata* fruits beneath
    the parent tree: predispersal attack by bruchids. Principes
    15:89-101.
Janzen, D. H. 1971c. Escape of juvenile *Dioclea megacarpa*
    (Leguminosae) vines from predators in a deciduous tropical
    forest. Amer. Natur. 105:97-112.
Janzen, D. H. 1971d. Escape of *Cassia grandis* L. beans from
    predators in time and space. Ecology 52:964-979.
Janzen, D. H. 1972. Escape in space by *Sterculia apetala* seeds
    from the bug *Dysdercus fasciatus* in a Costa Rican deciduous
    forest. Ecology 53:350-361.
Janzen, D. H. 1973a. Community structure of secondary compounds
    in plants. Pure Appl. Chem. 34:529-538.
Janzen, D. H. 1973b. Comments on host-specificity of tropical
    herbivores and its relevance to species richness. p. 201-211.
    In V. H. Heywood, ed. Taxonomy and ecology. Syst. Assoc.
    Special Vol. No. 5. Academic Press, London.
Janzen, D. H. 1973c. Sweep samples of tropical foliage insects:
    effects of seasons, vegetation types, elevation, time of day,
    and insularity. Ecology 54:687-708.
Janzen, D. H. 1973d. Tropical agroecosystems. These habitats
    are misunderstood by the temperate zones, mismanaged by the
    tropics. Science 182:1212-1219.
Janzen, D. H. 1974a. Swollen-thorn acacias of Central America.
    Smithson. Contrib. Bot. No. 13. 131 p.
Janzen, D. H. 1974b. The role of the seed predator guild in a
    tropical deciduous forest, with some reflections on tropical
    biological control. p. 3-14. *In* D. Price Jones and M. E.
    Solomon, eds. Biology in Pest and Disease Control. Blackwell
    Sci. Pub., Oxford.
Janzen, D. H. 1974c. The deflowering of Central America. Natur.
    Hist. 83:48-53.
Janzen, D. H. 1975a. Sweep samples of tropical deciduous forest
    foliage-inhabiting insects: seasonal changes and inter-field
    differences in adult bugs and beetles. Ecology. In press.
Janzen, D. H. 1975b. Intra- and inter-habitat variation in *Guazuma
    ulmifolia* (Sterculiaceae) seed predation by *Amblycerus ciste-
    linus* (Bruchidae) in Costa Rica. Ecology. In press.

Janzen, D. H., M. Atarroff, M. Fariñas, S. Reyes, N. Rincon, A. Soler, P. Soriano, and M. Vera. 1975. Changes in the arthropod community along an elevational transect in the Venezuelan Andes. In press.

Janzen, D. H., and C. M. Pond. 1975. A comparison, by sweep sampling, of the arthropod fauna of secondary vegetation in Michigan, England and Costa Rica. Trans. R. Entomol. Soc. Lond. In press.

Johnson, C. D. 1973. A new *Acanthoscelides* from *Indigofera* (Coleoptera: Bruchidae). Col. Bull. 27:169-174.

Johnson, R. M., and W. D. Raymond. 1964. The chemical composition of some tropical food plants: II. Pigeon peas and cow peas. Trop. Sci. 6:68-73.

Liener, I. E. 1974. Phytohaemagglutinins: their nutritional significance. J. Agr. Food Chem. 22:17-22.

Long, A., and P. S. Martin. 1974. Death of American ground sloths. Science 186:638-640.

Magnus, K. E., and C. E. Seaforth. 1965. *Samanea saman* Merrill: the rain tree. A review. Trop. Sci. 7:6-11.

Rehr, S. S., P. P. Feeny, and D. H. Janzen. 1973a. Chemical defence in Central American non-ant-acacias. J. Anim. Ecol. 42:405-416.

Rehr, S. S., D. H. Janzen, and P. P. Feeny. 1973b. L-Dopa in legume seeds: a chemical barrier to insect attack. Science 181:81-82.

Rehr, S. S., E. A. Bell, D. H. Janzen, and P. P. Feeny. 1973c. Insecticidal amino acids in legume seeds. Biochem. Syst. 1:63-67.

Rosenthal, G. A. 1972. Investigations of canavanine biochemistry in the jack bean plant, *Canavalia ensiformis* (L.) DC. II. Canavanine biosynthesis in the developing plant. Plant Physiol. 50:328-331.

Sevilla-Eusebio, J., J. C. Mamaril, J. A. Eusebio, and R. R. Gonzales. 1968. Studies on Philippine leguminous seeds as protein foods. 1. Evaluation of protein quality in some local beans based on their amino acid patterns. Philipp. Agric. 52:211-217.

Wilson, D., and D. H. Janzen. 1972. Predation on *Scheelea* palm seeds by bruchid beetles: seed density and distance from the parent palm. Ecology 53:954-959.

SYMPATRIC SPECIATION IN PHYTOPHAGOUS PARASITIC INSECTS

Guy L. Bush

Department of Zoology, University of Texas

Austin, Texas 78712, U.S.A.

The appearance of new insect pests on economically important plants is a well-known phenomenon to many applied biologists. In addition, populations of introduced or native insects are frequently encountered which exhibit different host preferences, but which are morphologically indistinguishable from one another (Brues 1924, Simms 1931, Mayr 1942, Andrewartha and Birch 1954, Zwölfer and Harris 1971). These so-called host races sometimes actually represent previously unrecognized reproductively isolated sibling species. Others appear to retain their distinct host preferences and other biological traits in the absence of any observable barriers to gene flow between the races. Two classic examples in North America are the codling moth (*Laspeyresia pomonella*), introduced from Europe in 1750, which shifted from apples to walnuts about 26 years after it reached California in 1873 (Essig 1931, Foster 1912), and the apple maggot (*Rhagoletis pomonella*) which moved from its native host hawthorn to introduced apples in 1864 and cherries less than 20 years ago (Bush 1966, 1969a,b, 1974).

It is now widely accepted that the vast majority, if not all, animal species including plant feeding parasitic insects speciate allopatrically only after a "genetic revolution" has occurred between two or more populations during extended periods of physical (i.e., geographic) isolation (Mayr 1954, 1963). However, the well documented appearance in recent times of new host races which acquire the characteristics of sibling species under what appear to be entirely sympatric conditions makes such an interpretation questionable. The origin and evolution of these host races and sibling species, therefore, remains an unresolved and controversial subject to some evolutionary biologists.

The problem of speciation in phytophagous parasitic insects is not only of academic interest but also a matter of considerable economic importance as well. Knowledge of the evolution of new pests capable of devastating important food and fiber crops is essential for the development of realistic biological control and quarantine measures (i.e., genetic manipulation, sterile insect releases, plant resistance, etc.). Understanding the genetic basis of rapid host race formation would also be essential for understanding and modeling aspects of community ecology of insect pest populations and insect-host plant interactions.

The development of a realistic model of speciation in para- sitic insects has been hampered in the past by a lack of informa- tion on certain critical features of their behavior, ecology, and genetics. Therefore, it has been relatively easy for the propo- nents of allopatric speciation to dismiss reputed cases of sympatric speciation because they were supposedly founded on biologically untenable assumptions and were unnecessarily compli- cated. Mayr (1947, 1963), for instance, lists several objections to the hypothesis of sympatric speciation. He argued that most proponents erroneously assume strong homogamy, a linkage between mate preference and habitat preference, conditioning (associative learning) without isolation and preadaptation to a habitat. The genetic mechanisms underlying these attributes and the effects of dispersal on continuously mixing populations he claimed were not fully appreciated.

However, lacking crucial facts about these attributes in parasitic insects, Mayr based his arguments almost without exception on evidence derived from studies on non-parasitic animals such as birds, fish, mammals, and certain insects like *Drosophila*. These animals have quite different adaptive strategies of niche selection, mating behavior, competition, associative learning patterns, and dispersal from those found in parasitic animals. This so-called evidence has now been accepted as representative of animals and incorporated into a body of conventional wisdom by most evolutionary biologists who now regard the problem resolved in favor of the allopatric model of speciation.

The lack of information on parasitic insects stems from the fact that most studies on these animals have touched on only one or two aspects of their biology; the emphasis was placed on the solu- tion of applied rather than evolutionary problems. For years investigations into the relationship between the Hessian fly (*Heteroderma rostochiensis*) and wheat (*Triticum*) concentrated on the genetic aspects of resistance in the host plant to attack by the fly. It was not until the recent excellent genetic studies

by Hatchett and Gallun (1970) on the gene-for-gene associations of
Hessian fly host races with various cultivars of wheat that the
genetics of survival from the parasites' point of view has been
given much attention.

However, survival is only one factor involved in host shifts.
The genetic bases for such attributes as host selection, diapause,
activity rhythms, dispersal, associative learning and other biological factors important to the problems of adaptation and host
race formation in the Hessian fly and other parasites have been
almost completely ignored. Frequently even a rudimentary knowledge of a parasite's ecology, mating behavior, and host range
is lacking. For example, the success of the sterile insect method
is ultimately dependent upon mating success, yet the mating behaviour of a pest as important as the screwworm fly in nature is
still unknown.

Because there were too many unanswered questions about the
origin and evolution of parasitic insects, I undertook a broad
study of a group of phytophagous parasitic flies of the family
Tephritidae. Several species in the genus *Rhagoletis* had recently
established new host races on introduced commercial fruits and
therefore were excellent subjects for an investigation of speciation
mechanisms. It is on this group that my colleagues and I have
concentrated our efforts and on which the present discussion is
based.

SOME IMPORTANT ASPECTS OF *RHAGOLETIS* BIOLOGY

An outline of important aspects of host and mate selection
in *Rhagoletis* is presented in Fig. 1. Additional details and
references to earlier work on the various steps in Fig. 1 can
be found in Bush (1974). This distilled picture represents the
combined efforts of several outstanding biologists including
Dr. Ronald Prokopy who is responsible for much of the information
on behavior, Dr. Ernst Boller who has provided important ecological
data and my graduate students, Dr. Milton Huettel and Mr. Stewart
Berlocher who have contributed much needed insight into the
genetics of tephritid populations.

The combined process of host and mate selection in the sexually mature apple maggot fly begins, 1, with the fly visually
selecting an object which is more or less the shape and color of
a tree or large shrub. This attraction to physical cues is in
no way host specific. The fly will move toward any plant approximately the right shape, size and color. Final detection of the
correct host is made on the basis of olfactory and contact chemical
cues, 2, emanating from the fruit and leaves of the tree. Fruit is

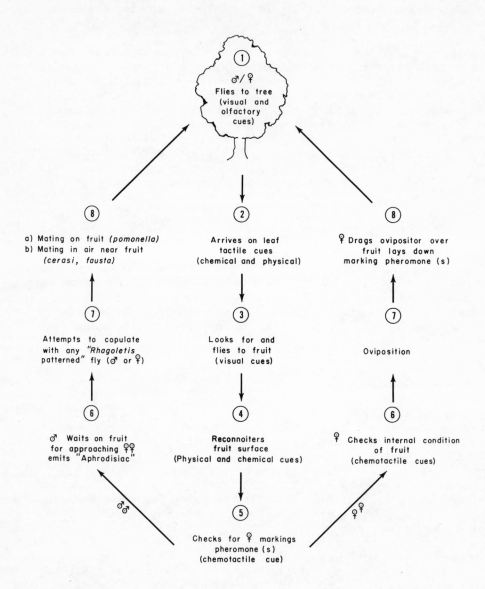

Fig. 1. Host selection and mating behavior in *Rhagoletis* (modified from Bush 1974).

selected for oviposition by the female and courtship by the male
on the basis of shape, size, color, 3, texture and apparently by
both external, and in the case of the female internal, chemical
cues, 4; these latter cues have not been identified as yet.
Specific host discrimination is therefore determined by chemical
rather than physical cues, a feature common to many phytophagous
insects (Dethier 1954).

These chemical cues are perceived by various chemoreceptors
whose specific sensitivity is dependent upon a receptor protein
(Ferkovich and Norris 1971). Host recognition is therefore ultimately dependent on the fly's genotype which encodes the amino
acid sequence of the receptor protein. Thus, a mutation altering
the amino acid composition of a receptor protein could have a
profound effect on the chemoreceptor's ability to recognize a
specific host plant chemical. Mutations could also affect the
way the central nervous system decodes all incoming sensory
information, thereby altering host odor or taste perception.

Courtship and mating occur on the host fruit 5-8 in
Rhagoletis and involve the use of both visual signals and pheromones, yet these important factors cannot be regarded as the most
significant isolating mechanisms between these closely related
species that court and mate on different species of plants. The
critical isolating mechanisms are ecological not ethological.
The fact that mating occurs only on the host plant in *Rhagoletis*
has led to some interesting patterns of evolution in these flies
(Bush 1969a,b).

PATTERNS OF EVOLUTION

In some species-groups, such as the *pomonella* group (Table 1),
speciation has always been accompanied or preceded by a shift to
a new host plant. Generally, these groups consist of several
sympatric sibling species with little or no morphological differentiation. Sibling species of *Rhagoletis* and other Tephritidae
which share the same host plant have only been encountered in
nature when they are geographically isolated from one another
(e.g., *R. indifferens* and *R. cingulata*) (Bush 1966, 1969a).

In other species-groups, speciation has not involved a shift
to a new host. For instance, the five members of the *suavis* species
group infest only walnuts (*Juglans*) (Table 1). At least four
of the species are sympatric over part or all of their ranges and
can sometimes be found at the same time ovipositing on fruits of
the same tree. As mating occurs on the host fruit, encounters
between species occur commonly. It is not surprising, therefore,
that all five species differ from one another in wing pattern and

Table 1.   The host of three major *Rhagoletis* species groups
           (modified from Bush 1969b).

Species Group	Host Plant Family	Adults Reared From	No. Plant Spp. Infested
POMONELLA GROUP			
R. pomonella	Rosaceae	*Crataegus*	15
		Pyrus	3
		Cotoneaster	1
		Prunus	3
R. mendax	Ericacae	*Vaccinium*	9
		Gaylussacia	3
R. cornivora	Cornaceae	*Cornus*	3
R. zephyria	Caprifoliaceae	*Symphoricarpus*	1
R. n. sp. A	Cornaceae	*Cornus*	1
CINGULATA GROUP			
R. cingulata	Rosaceae	*Prunus*	5
R. indifferens	Rosaceae	*Prunus*	5
R. osmanthi	Oleaceae	*Osmanthus*	1
R. chionanthi	Oleaceae	*Chionanthus*	1
SUAVIS GROUP			
R. suavis			
R. completa			
R. juglandis	Juglandaceae	*Juglans*	8
R. boycei			
R. zoqui			

body coloration and that three of the five species are sexually
dimorphic for these characters.   Selection has enhanced the
species specific visual cues used in the courtship and mating
which reduce the chances of interspecific mating.

Why some species groups have a propensity for shifting to a
new host plant in the course of speciation while others appear to
be irreversably tied to one host plant is not clear.   Our genetic
studies using polymorphic enzymatic and nonenzymatic proteins
suggest that this property might in some way be related to the
level of genetic polymorphism.   In "host shifters" there is a
substantially higher level of genetic polymorphism than present
in groups restricted to a single host plant (Berlocher and Bush,
unpublished).   It is likely that the "host shifters" are also more
genetically variable at loci that may be involved in a shift to a
new host.

THE GENETICS OF HOST SHIFTS

There are at least two major genetic components involved in host shifts by members of the genus *Rhagoletis*. The first includes the loci controlling host recognition and selection by the adult and the second is represented by those genes involved with survival in the larvae. As in most truly parasitic insects, the larva has no choice but to feed on the plant or animal selected by the adult female for oviposition. If the female errs in host recognition and selects a host in which the larva cannot survive then her progeny are lost. It is for this reason that a detailed knowledge of both adult and larval biology and genetics is so important.

Mutations that alter the response of a chemoreceptor or the decoding pathways in the central nervous system are the most important to consider in host shifts because specific host information is provided by chemical cues. What little information is available about the genetics of host selection in insects has been reviewed by Huettel and Bush (1971). We also presented evidence of their own that host selection is controlled by a single major locus in two closely related species of Tephritidae belonging to the genus *Procecidochares* which form galls on Compositae. Through a series of hybridization and backcross experiments we were able to demonstrate that individuals of *P. australis* infesting *Heterotheca* are homozygous for a pair of host selection alleles and oviposit only on *Heterotheca* when offered a choice of host and non-host plants. The sibling species *P. actitis* oviposits only on *Macroanthera*.

Attempts to study the genetics of host selection by interspecific hybridization experiments on other insect species, such as in diprionid wasps (Knerer and Atwood 1973), have involved an analysis of only F_1 or F_2 progeny and the number of genes involved has never been established. As far as I have been able to ascertain, no one has established if genetic polymorphisms for host selection exist *within* any natural population of a parasitic insect. This is obviously an area that needs further research and one that has important implications in the control of pest species.

The second class of genes that are important in some host shifts include those loci involved with survival of the larvae in the new host. I say some because it is clear that certain host fruits probably have sufficient nutrient qualities to maintain larval life for most *Rhagoletis* species. It is, therefore, possible to rear larvae of several *Rhagoletis* species on fruits rarely if ever infested under natural field conditions (Lathrop and Nickels 1932, Pickett and Neary 1940, Boller, unpublished).

The fruits of some plants, however, have various secondary plant compounds which inhibit feeding or which are actually lethal (Glasgow 1933). The highly toxic phenolic compound juglone for instance is present in high concentration in walnuts infested by members of the *suavis* group, and the solanaceous fruits attacked by some *Rhagoletis* species such as *R. striatella* and *R. lycopersella* contain alkaloids of varying toxicity. The ability to overcome these chemical defense mechanisms by an invading insect is dependent either directly or indirectly upon the action of specific enzymes.

Knerer and Atwood (1972, 1973) have indicated that their hybridization experiments suggest a polygenic control of survival in diprionid sawflies although they present no data. The inter-specific crosses carried out by Huettel and Bush (1971) during their studies on the genetics of host selection in *Procecidochares* support this conclusion. In forced oviposition experiments larvae of both species failed to develop from eggs deposited on non-host plants infested by other *Procecidochares* species. F_1 progeny, however, could survive on both host plants.

In insects there is only one well documented case of the genetics of survival genes involving host races. Hatchett and Gallun (1970) demonstrated that in the parasite-host system of the Hessian fly and wheat a gene-for-gene relationship exists between host resistance and parasite survival. Resistance in each cultivar of wheat is usually conferred by a dominant gene, while the ability to overcome the resistance by a race of the Hessian fly is controlled by a recessive "survival" gene. Such gene-for-gene relationships are commonplace in races of pathogenic fungi and have also been reported in nemetodes, bacteria, viruses and higher plant parasites (Day 1974).

The total number of genetic changes occurring during a host shift that has been accompanied by speciation has yet to be established for any insect. Lewontin (1974) and others have sug-gested that species differ in at least 10% of their genome. He regards this as a lower limit. His estimate is based on studies of allozyme variation in sibling species and geographic races of *Drosophila* which have been isolated for long periods of time. The experimental studies on phytophagous parasites and on at least one *Drosophila* (Prakash 1972) indicates that in some cases the number of genes involved in speciation events may be few in number and far less than the minimal 10% proposed by Lewontin. This would be particularly true if the shift was between plants of similar chemistry, which is frequently the case.

Two other classes of genes could be important in host shifts in speciation: (1) those which determine to what degree the insect can be induced to recognize one host plant from another and (2) the genetic component that regulates diapause. The larvae of *Manduca sexta* and *Heliothis zea* for instance can be induced to show a clear preference for the plant previously eaten. However,

no induction is possible on plants outside the insect's normal
host range (Jermy et al. 1968).

Huettel and Bush (1971) also found evidence that larval conditioning had a strong effect on host plant selection. Hybridization experiments suggest that induction is probably under polygenic
control. However, the capacity to be induced to select a host
plant for oviposition or mating appears to be ultimately dependent
upon the ability of the chemoreceptors to respond to a specific
host plant cue. An individual may be genetically programmed to
prefer one host but be incapable of perceiving it as its chemoreceptors are not sensitive to the chemical cue emitted by the
plant.

Those genes regulating the termination of diapause probably
play an important role in increasing the level of reproductive
isolation between the host races of temperate climate species after
the initial host shift has successfully been made. Fig. 2 represents a graphic picture of the approximate allochronic emergence
patterns of the three races of *R. pomonella* in Door County,
Wisconsin. Although little or no overlap apparently exists in the
emergence of the cherry race with that of the hawthorn race, late
emerging members of the apple race overlap with early emerging
members of the hawthorn population. Allochronic isolation has
also been important in the host race formation and speciation
process of the *cingulata* species group in North America whose
members infest cherries and olives (Table 1).

The western cherry fruit fly, *R. indifferens*, in California
normally infests the native bitter cherry, *Prunus emarginata*, in
August. This plant grows at altitudes above 3,500 feet and fruits
late in the summer and fall (Fig. 3). The sweet cherry, *P. avium*,
and sour cherry, *P. cerasus*, although introduced to California over
a hundred years ago are normally not attacked by the fly. Occasionally, however, a single domesticated cherry orchard growing within
the upper limit of their altitudinal range near the wild bitter
cherry will become infested. These populations are routinely
eradicated by the California Department of Agriculture. Usually
the same orchard is free from attack the following year.

To the north (Oregon, Washington, and British Columbia) the
fly established permanent populations on domestic cherries shortly
before 1913, some 89 years after cherries were introduced to the
Northwest. Distinct races now appear to coexist in these areas,
one infesting domestic cherries at low altitudes during late June
and early July and the other attacking the native bitter cherry
at high altitude during late July and August. As far as I have
been able to ascertain, the two *indifferens* races are almost completely allochronically isolated from one another north of California

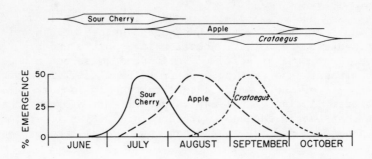

Fig. 2. Approximate allochronic emergence patterns of *Rhagoletis pomonella* in Door County, Wisconsin.

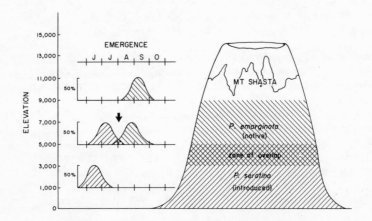

Fig. 3. Interrelationship between temporal and spatial factors in the shift of *Rhagoletis indifferens* from the bitter cherry to domestic cherry. Shifts can only occur at around 5,000 feet during the last two weeks of July.

(Bush 1966, 1969b). Furthermore, in some areas like the Okanagan
Valley of British Columbia the wild host may be uninfested while
the earlier ripening domestic cherries are heavily infested by
the fly (Madsen and Arrand 1970). Possibly conditions in the late
summer and fall are too adverse for the fly to complete development in the bitter cherry.

Based on our knowledge of what has been occurring in recent
years in the California populations of *R. indifferens* we can
tentatively reconstruct how the shift from wild to domestic cherries
occurred in Oregon in 1913. The situation in California also provides us with an excellent example of how environmental conditions
affect host shifts and speciation in other phytophagous parasitic
insects.

In California there is only a narrow window in space and time
when such factors as fly emergence, fruit maturation and abundance,
and the flies' genetic constitution are such that a successful host
shift can occur. For reasons outlined previously (Bush 1974),
successful host shifts are most likely to occur when the two host
plant species coexist sympatrically and with flies that have a
suitable preadapted genotype for coping with the new host fruit.
On Mt. Shasta, for instance, the shift can occur only in the narrow
zone at between 3,500 to 5,000 ft. during the last two weeks of
July (Fig. 3). If the new race was permitted to become established
in the domestic cherry it would eventually expand its population
by simultaneously shifting its emergence period to an earlier date
and moving to lower altitudes where greater numbers of domestic
cherries are grown. Eventually two populations would coexist which
would have different emergence times and host preferences and for
the most part overlap geographically in only a very small area.

Allochronic isolation is even more fully developed between
two sibling species in the *cingulata* group both of which infest
native olives. *R. chionanthi* infest the fruits of the fringe-tree,
Chionanthus virgincus, which fruits in the summer, while the larvae
of *R. osmanthi* are found in devilwood, *Osmanthus americanus*, which
fruits during midwinter (Bush 1969). As illustrated in Fig. 4,
the emergence times of the two fruit fly species probably diverged
as *Osmanthus* shifted its fruiting time to cooler winter months in
response to climatic changes occurring during the Pleistocene.

Similar differences in emergence patterns have been observed
between sibling species within the *pomonella* sibling species group
(i.e., *pomonella* and *mendax*) (Lathrop and Nickels 1932) and between
the European cherry fruit fly (*Rhagoletis cerasi*) infesting *Prunus*
and sympatric populations of an undescribed sibling species feeding
in honeysuckle (*Lonicera*) (Boller and Bush 1973).

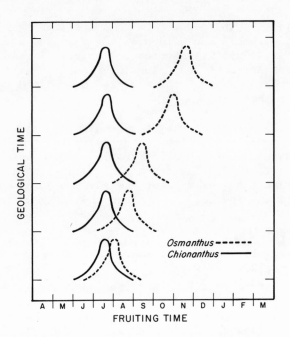

Fig. 4. The probable origin of present day allochronic isolation between *Rhogletis chionanthi* infesting *Chionanthus* in the summer and *R. osmanthi* that infests *Osmanthus* in the winter by a shift in fruiting time of *Osmanthus*.

The genetic basis for these differences in diapause is unknown, and in fact has been studied in only a few insect species. In all cases diapause appears to be controlled by two or more loci and in the silkworm six genes controlling egg diapause have been identified (Beck 1968). Studies attempted thus far have concentrated on differences in diapause patterns between geographic races of the same species. How many genetic differences exist between closely related host races or sibling species is unknown but they are probably also polygenic in nature.

Undoubtedly a shift from one host fruit with a late summer maturation period to another host which fruits in mid or early summer would have a profound effect on which diapause alleles would be favored in the new host race. An initial shift may be made by early emerging flies reared on the original host ovipositing on late maturing fruit of the new host. At least two individuals, a male and female, would have to make the shift as mating normally occurs after dispersal when the adults are over eight days old. These flies would possess genotypes for early emergence and produce progeny with similar early emerging genotypes. Selection in

subsequent generations infesting the new host plants would further
tune the genotype to insure that emergence time was entrained to
the ripening period of a new host plant.

There are undoubtedly other genetic loci that might undergo
adaptive changes over a period of time following the initial colonization which would allow the fly to better exploit the new host.
Enzymes involved with digestion and flight would be operating under
different temperature regimes and in some cases on different sub-
strates. Changes at these loci might alter the fitness of alleles
and other loci resulting in the genetic alteration of the genome
of the new host race. Selection would eventually lead to a re-
ordering of at least a portion of the genome into a new coadapted
gene pool in the new host race. Such changes would only further
enhance the degree of reproductive isolation between the old and
new populations with the passage of time.

A GENERALIZED GENETIC MODEL OF SYMPATRIC SPECIATION

It is now quite clear that host races of phytophagous parasitic
insects have evolved sympatrically. Furthermore these host races
are undoubtedly the progenitors of the many reproductively isolated sibling species so frequently found coexisting sympatrically
on different host plants. Although sympatric speciation is
probably a widespread phenomenon in stenophagous parasitic insects
(see for instance Knerer and Atwood 1973, Zwölfer and Harris 1971),
only certain groups are preadapted to this mode of speciation. In
the case of phytophagous parasitic insects, they usually possess
the following biological traits:
1. Mating occurs on or near the host plant
2. The adult female selects the host; the larvae have no
    choice
3. They are monophagous or stenophagous, attacking groups of
    closely related host plant species
4. Host selection is under genetic control
5. Larval survival is dependent upon the action of survival
    genes

Other factors could act to reinforce reproductive isolation
and probably hasten the development of new species. These include:
6. Univoltinism with genetic control of emergence time
7. Genetic control of host plant induction (conditioning)

Host selection and survival genes are obviously the two most
important genetic factors that must be incorporated into any realistic genetic model of sympatric speciation. The most critical
discussion of sympatric speciation from a theoretical and mathematical point of view is that of Maynard Smith (1966) who was able

to demonstrate that a "stable polymorphism between alleles adapting individuals to different ecological niches could be the first stage in sympatric speciation." Furthermore, he was able to show that if such a polymorphism was accompanied by another polymorphism that causes positive assortative mating (i.e., flies best adapted to a particular niche tend to mate with one another) and if there is some degree of habitat selection, then two reproductively isolated populations would evolve.

In many phytophagous parasitic insects such as Chrysomilidae, Curculionidae, diprionid sawflies, Agromyzidae and Tephritidae (Bush 1969a, Knerer and Atwood 1973, Nowakowski 1962, Zwölfer and Harris 1971), mating occurs on the host which serves as both a rendezvous for courtship and mating as well as food for the immature stages. Therefore a polymorphic host selection gene such as the one found in *Procecidochares* could also serve as Maynard Smith's assortative mating gene. When this locus is coupled with a polymorphic survival gene in a new host race, mating with other individuals with similar genotypes would occur on the host for which they are best adapted.

In a population of fruit flies polymorphic for an allele pair S_1, S_2 where each allele confers the ability to survive in one host plant and which is also polymorphic for a second allele pair H_1, H_2 causing positive assortative mating (H_1H_1 selects host A, H_2H_2 selects host B with mating occurring on the host plant), then two reproductively isolated populations will evolve. Fig. 5 reconstructs the course of events leading to sympatric speciation. The example is based on the shift of *R. pomonella* from apples to cherries in Door County, Wisconsin.

A host shift is initiated by mutations which introduce the new host selection ($H_1 \rightarrow H_2$) and survival ($S_1 \rightarrow S_2$) alleles into the apple-infesting population ($H_1H_1 S_1S_1$). These new alleles could be retained in the population at low level by chance, because they are closely linked to a beneficial gene or through over-dominance. Certain recombinants would be functionally lethal. All H_1H_1 and H_1H_2 females would oviposit on apple, but any S_2S_2 larvae would not survive. Likewise, H_2H_2 females would oviposit on cherries, but their S_1S_1 progeny would die on cherries.

Individuals heterozygous for the host selection locus H_1H_2 could potentially oviposit on both plants. Induction and allochronic isolation, however, would inhibit random mating and oviposition between the two races. Most of the heterozygotes would remain on apple where coadapted induction genes would be strongly developed and operate most efficiently. Only the few H_2H_2 flies produced each year early in the season would seek out the cherry for mating and oviposition.

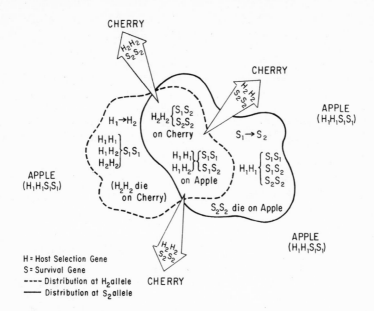

Fig. 5. A graphic model reconstructing the sympatric origin at the genetic level of the cherry race of *Rhagoletis pomonella* from the apple race in Door County, Wisconsin.

Gene flow would be further inhibited as a result of a shift to an earlier emergence period by the cherry race once it was established. Not only would the H_2H_2 flies be mating and ovipositing on the new host, but those individuals emerging early would find a greater abundance of oviposition sites. Selection would therefore favor individuals with even earlier emerging genotypes resulting in a shift in emergence times and increased allochronic isolation between the apple and cherry races. In the early stages of development the new host race would also escape from its parasites and some predators.

Once established, the cherry race would rapidly increase in numbers and spread to areas beyond the region where the original mutations arose in the apple race. In this original site a small amount of gene flow might occur each year as homozygous H_2H_2 individuals shift from apple to cherry trees. The effect of this gene flow on the speciation process would probably be negligible, however. Selection would immediately act on a constellation of genes to better adapt the new race to cherry once a population was established. With each successive generation not only would the numbers of individuals on cherry increase, but a stronger coadapted gene pool would evolve. The few flies moving from apples to cherries would find their genome increasingly ill adapted and

eventually eliminated altogether. The end result would be complete
reproductive isolation between the old and new races.

If this scenario of host race formation and speciation is true
then new host races may be established by more than a single founder.
One would therefore expect to find little evidence for a major
genetic revolution between newly established host races. Our
investigation of allozyme differences in recently established
host races of *Rhagoletis*, as well as several sibling species which
are at least over ten thousand years old, has confirmed that no
genetic revolution appears to accompany the establishments of new
races or sibling species (Bush 1974, and in preparation). The
process of coadaptation mentioned earlier apparently involves only
a small portion of the genome.

As I have pointed out before (Bush 1974), shifts from a wild
to a domesticated host or from one domesticated host to another,
as in the case of *R. pomonella*, probably occur much more frequently
than might be expected between two wild host plants under natural
conditions. Domesticated plants frequently have been carefully
selected to reduce many of the chemical agents, or allomones
(Whittaker and Feeny 1971), which protect the plant from attack
by herbivores and parasites. Furthermore, inbreeding, grafting
and other vegetative methods of propagation have greatly reduced
the amount of genetic variability between individuals within a
domesticated species. Whole areas of a country may be planted with
individuals of the same genotype and the results can be disastrous.
Over 12% of the corn crop in the United States was lost several
years ago when a new race of the corn blight invaded seed corn
varieties bearing identical cytoplasms (Tatum 1971).

Shifts between or onto domesticated hosts may therefore require
fewer alterations of the genome than shifts occurring between wild
species of host plants. Successful shifts are also more likely
to occur between host species with similar chemical constitutents
than between plants which are biochemically markedly different.
One would predict that within a species group where host shifts
have accompanied speciation, the chemistry of the various plants is
either very similar or the shifts have occurred between plants
belonging to the same genus or family of plants. This is exactly
the pattern encountered in *Rhagoletis*.

Even where shifts occur between very different hosts, such
as in the case of the codling moth which shifted from apples to
walnuts, the plants may not be as different to the parasite at the
chemical level as they appear to be to us. Obviously, a concerted
effort is needed to establish what properties of the hosts are
actually critical to larval survival and adult host recognition.
Much more information is also needed on genetic aspects of host

selection, of survival, associative learning behaviour and diapause and the role they play in the determination of the natural host range of parasites.

The proposed model does not exclude the possibility that sympatric speciation in phytophagous parasitic insects might occur by some other means. Although the reproductive isolating mechanisms clearly implicated in host race formation are those involved in habitat selection (i.e., host preference and survival, seasonal isolation, and associative learning or induction), other premating and postmating mechanisms may be involved.

The possible role in sympatric speciation through changes in ethologically-based reproductive isolating mechanisms, such as the type recently uncovered in the pheromone races of the corn borer (Kochansky et al. 1974), should be explored. Chromosomal rearrangements (White 1973) and symbionts (Dobzhansky 1972) have also been implicated in the parapatric or sympatric speciation and rapid evolution of other insects and higher animals. The importance of these factors in the speciation of parasitic insects remains unknown. It is not entirely inconceivable that some of these factors acting in consort with a genetically based host shift might further enhance the level of reproductive isolation between host races and accelerate the speciation process.

One question concerning sympatric speciation which I have purposely avoided until now is: "How does one determine when speciation has occurred in a parasite?" Some would probably regard the initial establishment of a new host race as the critical point of speciation in which case speciation would occur in a matter of one or two generations. Some gene flow might continue for many years between the old and new races but have little effect on the final development of complete reproductive isolation. Others might prefer to postpone the confirmation of species status to that moment in time when gene flow is completely eliminated.

In the long run, the decision to elevate a host race to the status of a distinct species is somewhat arbitrary. The transition from host race to species at the genetic level may involve the alteration of only a few genes or many depending on the situation. There is probably no single generation that can be unequivocally identified as the one in which speciation occurred. Whether a population is designated a host race or species must often rest on the rather subjective opinion of those most familiar with the species in question. Because the transition phase is dependent on so many intrinsic and extrinsic variables, each case must be interpreted on its individual merits and it is unlikely that the same pattern of host race formation and speciation applies to all parasitic insects.

It should also be pointed out that attempts at colonization of
a host plant probably occur much more frequently than we now realize.
The vast majority of the beachheads established on these new host
"islands" are probably doomed to failure either by chance extinction,
competition, insufficient genetic diversity, or a multitude of other
factors. Only rarely will a new host race expand its beachhead and
establish a permanent population. The total number of speciation
events is therefore likely to be much greater than the actual observed number of successfully competing species of parasitic insects.

SUMMARY

There is now little doubt that sympatric speciation has played
an important role in the evolution of phytophagous parasitic insects. The establishment of new host races, particularly on chemically related or domesticated plants, may initially require only
minor alterations in the genome. Genetic changes which alter host
selection behavior and survival can erect strong barriers to gene
flow between the parent population on the original host and the new
host race. The alteration of genetic factors regulating emergence
patterns (diapause) and induction (associative learning) behavior
can reinforce reproductive isolation and speed the process of
sympatric speciation.

The implication of sympatric host race formation and speciation
in parasites to man's present and future food needs is foreboding.
The green revolution and man's ever-increasing efforts to produce
more food can only further enhance the likelihood that more host
races will develop on domesticated plants. Such host races may
very likely transmit new diseases and compound the problems generated by our ever-increasing reliance on pesticides. Only by gain-
ing a better understanding of the mechanisms involved in host race
formation through basic studies on the ecology, behavior, and
genetics of parasites can we hope to reduce the likelihood of
complicating an already alarming situation.

ACKNOWLEDGMENTS

I would like to thank Stewart Berlocher, Craig Jordan, and
Richard Williams for their suggestions and critical review.

LITERATURE CITED

Andrewartha, H. G., and L. C. Birch. 1954. The distribution and
    abundance of animals. Univ. Chicago Press. 782 p. (p. 690-
    695).

Beck, S. D. 1968. Insect photoperiodism. Academic Press, New York. 288 p.

Boller, E. F., and G. L. Bush. 1973. Evidence for genetic variation in populations of the European cherry fruit fly, *Rhagoletis cerasi* (Diptera: Tephritidae) based on physiological parameters and hybridization experiments. Entomol. Exp. Appl. 17:279-293.

Brues, C. T. 1924. The specificity of food plants in the evolution of phytophagous insects. Amer. Natur. 58:127-144.

Bush, G. L. 1966. The taxonomy, cytology, and evolution of the genus *Rhagoletis* in North America (Diptera, Tephritidae). Bull. Mus. Comp. Zool. 134:431-562.

Bush, G. L. 1969a. Mating behavior, host specificity, and the ecological significance of sibling species in frugivorous flies of the genus *Rhagoletis* (Diptera, Tephritidae). Amer. Natur. 103:699-672.

Bush, G. L. 1969b. Sympatric host race formation and speciation in frugivorous flies of the genus *Rhagoletis* (Diptera, Tephritidae). Evolution 23:237-251.

Bush, G. L. 1974. The mechanism of sympatric host race formation in the true fruit flies (Tephritidae). p. 3-23. In M. J. D. White, ed. Genetic Analysis of Speciation Mechanisms. Aust. and New Zealand Book Co.

Day, P. R. 1974. Genetics of host-parasite interaction. W. H. Freeman and Co., San Francisco. 238 pp.

Dethier, V. G. 1954. Evolution of feeding preferences in phytophagous insects. Evolution 8:33-54.

Dobzhansky, Th. 1972. Species of *Drosophila*. New excitement in an old field. Science 177:664-669.

Essig, E. O. 1931. A history of entomology. Macmillan, New York. 1029 p. (p. 247-254).

Ferkovich, S. M., and D. M. Norris. 1971. Antennal proteins involved in the neural mechanism of quinone inhibition of insect feeding. Experientia 28:978-979.

Foster, S. W. 1912. On the nut-infesting habits of the codling moth. U.S. Dep. Agric. Bur. Entomol. Bull. 80:67-70.

Glasgow, H. 1933. The host relations of our cherry fruit flies. J. Econ. Entomol. 26:431-438.

Hatchett, J. H., and R. L. Gallun. 1970. Genetics of the ability of the Hessian fly (*Mayetiola destructor*) to survive on wheats having different genes for resistance. Ann. Entomol. Soc. Amer. 63:1400-1407.

Huettel, M. D., and G. L. Bush. 1971. The genetics of host selection and its bearing on sympatric speciation in *Procecidochares* (Diptera, Tephritidae). Entomol. Exp. Appl. 15:465-480.

Jermy, T., F. E. Hanson, and V.G. Dethier. 1968. Induction of specific food preferences in lepidopterous larvae. Entomol. Exp. Appl. 11:203-211.

Knerer, G., and G. E. Atwood. 1972. Evolutionary trends in the subsocial sawflies belonging to the *Neodiprion abietis* complex (Hymenoptera: Tenthredinoidea). Amer. Zool. 12:407-418.

Knerer, G., and G. E. Atwood. 1973. Diprionid sawflies: polymorphism and speciation. Science 179:1090-1099.

Kochansky, J., R. T. Cardé, J. Liebherr, and W. L. Roelofs. 1974. Sex pheromones of the European corn borer (*Ostrinia nubilalis*) in New York. Chem. Ecol. In press.

Lathrop, F. H., and C. B. Nickels. 1932. The biology and control of the blueberry maggot in Washington County, Maine. U.S. Dep. Agric. Tech. Bull. 275. 76 p.

Lewontin, R. C. 1974. The genetic basis of evolutionary change. Columbia Univ. Press, New York. 346 p.

Madsen, H. F., and J. C. Arrand. 1970. The biology and control of the cherry fruit flies in the Okanagen Valley of British Columbia. B. C. Dept. Agric. Bull. 70. 5 p.

Maynard Smith, J. 1966. Sympatric speciation. Amer. Natur. 100: 637-650.

Mayr, E. 1942. Systematics and the origin of species. Columbia Univ. Press. New York. 334 p. (p. 208).

Mayr, E. 1947. Ecological factors in speciation. Evolution 1: 263-288.

Mayr, E. 1954. Change of genetic environment and evolution. p. 157-180. In J. Huxley, A. C. Hardy, and E. B. Ford, eds. Evolution as a process. Allen and Unwin, London.

Mayr, E. 1963. Animal species and evolution. Harvard Univ. Press. Cambridge, Mass. 797 p.

Nowakowski, J. T. 1962. Introduction to a systematic revision of the family Agromyzidae (Diptera) with some remarks on host plant selection. Ann. Zool. Warsz. 20:67-183.

Pickett, A. D. and M. E. Neary. 1940. Further studies on *Rhagoletis pomonella* (Walsh). Sci. Agr. 20:551-556.

Prakash, S. 1972. Origin of reproductive isolation in the absence of apparent genetic differentiation in a geographic isolate of *Drosophila pseudoobscura*. Genetics 72:143-155.

Simms, A. D. 1931. Biological races and their significance in evolution. Ann. Appl. Biol. 18:404-452.

Tatum, L. A. 1971. The southern corn leaf blight epidemic. Science, 171:1113-1116.

White, M. J. D. 1973. Animal cytology and evolution. 3rd ed. Cambridge Univ. Press, London. 961 pp.

Whittaker, R. H., and P. P. Feeny. 1971. Allelochemics: Chemical interaction between species. Science 171:757-770.

Zwölfer, H., and P. Harris. 1971. Host specificity determination of insects for biological control of weeds. Annu. Rev. Entomol. 16:159-178.

AUTHOR INDEX

Alexander, R. D., 66, 84
Alloway, T. M., 25, 39
Anderson, W. W., 114, 128
Andrewartha, H. G., 187, 204
Arnaud, P. H., 103, 106, 109
Arrand, J. C., 197, 206
Arthur, A. P., 16, 25, 29, 30, 39, 42
Askew, R. R., 1, 5, 7, 11, 67, 73, 79, 84, 85, 98, 101, 131, 136, 140, 141, 142, 145, 149, 150, 152, 153
Assem, J. van den, 67, 68, 72, 73, 74, 75, 76, 77, 84
Ataroff, M., 181, 186
Atwood, G. E., 193, 194, 199, 200, 206

Baerwald, R. A., 73, 84
Bakker, K., 31, 39
Bandemer, S. L., 177, 184
Barass, R., 67, 77, 84
Barbehenn, K. R., 8, 11,
Barlow, J. S., 35, 39, 46
Barras, D. J., 33, 34, 35, 39, 47
Bartlett, B. R., 21, 39, 40
Batsch, W. W., 30, 39
Beaver, R. A., 101, 105
Beck, S. D., 198, 205
Bell, E. A., 155, 156, 173, 184, 186
Benskin, J. B., 35, 42
Beroza, M., 20, 22, 40, 43, 47
Berryman, A. A., 101, 105

Bess, H. A., 95, 101, 106, 109
Bierl, B. A., 20, 22, 43, 47
Birch, L. C., 187, 204
Birch, M. C., 20, 40
Blunck, H., 101, 107
Bobb, M. L., 101, 107
Boller, E. F., 197, 205
Borden, J. H., 23, 29, 44, 45
Bossert, W. H., 113, 128
Bousch, G. M., 73, 84
Bowman, M. C., 20, 22, 43
Bracken, G. K., 89, 101
Bradley, J. R., 30, 44
Brazzel, J. R., 26, 36, 43, 45
Brues, C. D., 187, 205
Buckingham, G. R., 73, 85
Bush, G. L., 187, 189, 190, 191, 192, 193, 194, 195, 197, 200, 202, 205

Callan, E. McC., 149, 152
Callegarini, C., 35, 44
Camors, F. B., 17, 18, 37, 40
Campbell, T. R., 79, 80, 85
Cardé, R. T., 203, 206
Carton, Y., 21, 40
Caudle, H. B., 16, 46
Center, T. D., 155, 173, 184
Chamberlin, F. S., 26, 40
Claridge, M. F., 149, 153
Clausen, C. P., 67, 85, 94, 95, 96, 98, 101, 110
Cleare, L. D., 101, 109
Cody, M. L., 114, 127
Cole, L. R., 73, 85
Collins, J. P., 50, 65

GENUS AND SPECIES INDEX

Note by Robert W. Matthews with reference to "Courtship in Parasitic Wasps", p. 66-86.

An extensive review of the literature by Gordh (1974) on courtship in parasitic Hymenoptera appeared after this paper was submitted. In particular Table 70 summarizes published records on courtship. Also included are extensive data on male sperm depletion in *Aphytis*.

Gordh, G. 1974. Biological and behavioral studies of the genus *Aphytis* and their taxonomic implications (Hymenoptera: Aphelinidae). Ph.D. Dissertation, Univ. of California, Riverside. 264 p.